CONTENTS

Preface

Introduction

1. **The Nature of a Psychology of Mathematics** 3
   A Psychology of Subject-Matter Learning 4
   Psychology and the Tasks of Instruction 6

PART I: MATHEMATICS AS COMPUTATION 9

2. **The Psychology of Drill and Practice** 11
   Edward L. Thorndike and the Formation of Bonds 12
   Drill Versus Meaningful Instruction 17
   What Makes Arithmetic Problems Easy or Hard? 19
   Optimizing the Effectiveness of Practice 24
   Drill and the Development of Automaticity 30
   Summary 35

3. **Transfer Hierarchies and the Organization of Instruction** 38
   Learning Hierarchies for Mathematical Tasks 39
   Introduction to Rational Task Analysis 57
   Summary 64

4. **Analyses of Performance on Computational Tasks** 67
   Simple Mathematical Tasks: The Use of Reaction-Time Data in Studying Performance 69
   Computational Strategies and Systematic Errors: Protocol Analysis 83
   Solving Context-Embedded Problems: Computer Simulation 89
   Summary 94

iv   CONTENTS

PART II: MATHEMATICS AS CONCEPTUAL UNDERSTANDING AND PROBLEM SOLVING   97

5. **Teaching the Structures of Mathematics**   101
   A Decade of Curriculum Reform   102
   Bruner and the Cognitive Representation of Mathematical Concepts   110
   Dienes' Multiple Embodiments and the Sequence of Instruction   116
   Questions Raised by the Structure-Oriented Approaches   123
   Summary   126

6. **Structure and Insight in Problem Solving**   128
   Gesalt Principles and Some Mathematical Examples   129
   The Process of Productive Thinking   138
   Implications of Gesalt Thinking for Instruction   146
   Summary   152

7. **Piaget and the Development of Cognitive Structures**   155
   Thinking as Structuring   156
   The Development of Piagetian Structures   165
   Critiques of Piaget   174
   Piaget and Instruction   186
   Summary   193

8. **Information-Processing Analyses of Understanding**   196
   The Organization of Semantic Memory   198
   Using Knowledge and Strategy for Solving Problems   213
   Strategies for Analyzing Problems and Searching Knowledge Structures   220
   Summary   235

CONCLUSION

9. **Looking Ahead**   241
   The Changing Nature of Instructional Psychology   242
   Needed: A Cognitive Learning Psychology   244
   Questions for the Psychology of Mathematics Instruction   246
   Applying Psychology To Instruction   252

AUTHOR INDEX   254

SUBJECT INDEX   257

Preface

This book is addressed to psychologists, educators, and mathematicians who are interested in the mental processes involved in learning mathematics. The book traces the history of psychologists' efforts to inform mathematics instruction, from the associationist work of Edward L. Thorndike to today's information-processing studies of mathematical thinking—with consideration of Gestalt, Piagetian, and various branches of American behavioral and cognitive psychology along the way. The book can thus be read as a history of psychologists' efforts to discover and explicate the nature of learning and thought processes in mathematics. But it is meant to be more than that. It is, above all, an effort to give shape and direction to an emerging branch of study concerned with how expert thought in mathematics proceeds, how that expertise develops, and how instruction can enhance the process of mathematics learning.

For many decades mathematicians and educators committed to improving the intellectual power of mathematics instruction were unable to find much of interest in the work of psychologists. This is not surprising, for psychologists—if they attended to mathematics at all—generally were attempting to make mathematical subject matter fit general laws of learning rather than trying to understand the processes of mathematical thought in particular. This is now changing. An emerging psychology of mathematics is focusing directly on the processes of mathematical thinking and on the ways in which people come to understand the structures of mathematics. This new line of investigation joins cognitive psychology's concern for the processes of thought with traditional learning psychology's interest in how new abilities are acquired. Increasingly it includes explicit attention to the role of instruction in the development of mathematical thinking. The groundwork is thus being laid for a theory of mathematics instruction rooted both in the structure of the subject matter and in the principles of cognition and learning.

In focusing on this new area of study we are expressing our hopes for the future rather than describing a fully developed line of research; for the new psychology of mathematics is still evolving. It is being created in the laboratories and the experimental classrooms of a few psychologists and educators, usually either highly trained in mathematics themselves or working in collaboration with mathematicians. It is gaining strength through formal and informal exchanges among psychologists, educators, and mathematicians. In the process, it is developing a point of view and some styles of research that are likely to distinguish it from past work in the psychology of mathematics learning. We attempt here to give a flavor of what is possible in this new domain of psychological investigation by providing many detailed examples of actual research efforts. We hope that by documenting the present state of the art we will help it take strength, form, and confidence in its own possibilities as a science.

We owe a great debt to the many people who graciously consented to read and comment on the manuscript at several stages in its development. William and Carol Rohwer, who were instrumental in getting the book under way, read very early drafts of several chapters and encouraged us to complete the rest. Solomon Asch, Jeffrey Bisanz, and Hans Furth contributed their expert knowledge to our interpretation of Gestalt and Piagetian psychology. Eugene Deskins, Rochel Gelman and Barbara Searle offered many helpful suggestions. Special thanks go to Mario Benedicty, James Greeno, and David Klahr, whose detailed comments on the entire manuscript helped shape the final version. The thoughtful critiques of all these people forced us to face difficult questions, but made the writing enterprise a most exciting challenge. Any remaining shortcomings in the book are, of course, solely our responsibility.

Major portions of the book were first written while one of us (LBR) was a Fellow at the Center for Advanced Study in the Behavioral Sciences, and we are grateful for the "head start" that that year of scholarly leisure gave to this venture. Throughout our collaboration, our colleagues at the Learning Research and Development Center of the University of Pittsburgh provided incentive and inspiration.

Valerie Shalin earns our thanks for her careful construction of the book's subject index. We are also pleased to acknowledge the skillful help of Donna Rottman in preparing the artwork for publication. For untold hours of manuscript typing, we thank Barbara Viccari. We are especially grateful to Cathlene Hardaway for her astute and meticulous assistance in the many details of technical production and editing. To our families, who quietly supported us in numerous ways throughout this endeavor, we offer an apology for letting it take so long.

Lauren B. Resnick
Wendy W. Ford

INTRODUCTION

1 The Nature of a Psychology of Mathematics

What could a book on the psychology of mathematics be about? Most people know psychology is concerned with how people learn, with how they perform tasks, and with how they develop. But few think of psychology as being concerned with special areas of interest such as mathematics, the social sciences, or any other subject taught in the schools. It is partly because people are puzzled by the notion of a psychology of mathematics that this book was written.

As psychologists concerned specifically with mathematics, our goal is to ask the same questions that experimental and developmental psychologists ask about learning, thinking, and intelligence but to focus these questions with respect to a particular subject matter. What this means is that instead of asking ourselves the general question, "How is it that people think?" we ask ourselves, "How is it that people think about mathematics?" Instead of asking "How do people's thought processes develop?", we ask, "How does understanding of mathematical concepts develop?" We want to know what mixture of experience and intellect makes that thing called *mathematical ability* happen. As psychologists of a subject matter, we want to know not simply how overall human performance becomes skillful but how human performance of mathematically significant skills becomes fluent, and how those skills are integrated in the context of mathematical problem solving.

When psychologists undertake to study complex tasks such as those in mathematics, they try to figure out what people actually do when they engage in these tasks, what behaviors constitute mathematics performance. Because mathematics is an intellectual activity rather than a physical one, psychologists must try to understand what is going on in people's heads. A group of students, for example, is presented with a page of new geometry problems. They struggle through them and at the end offer their answers. Sometimes their answers are right; sometimes they are not. Both cases are of interest to the psychologist, whose concern is with what has gone on in between, with what has happened between the posing of the problem and students' offering of an answer. The psychologist is also concerned with what kinds of experiences will help students perform well on problems of that type. How should the problems be presented? What sorts of practice might help? Are there particularly favorable times, in terms of students' intellectual development, to present certain kinds of material?

A PSYCHOLOGY OF SUBJECT-MATTER LEARNING

As psychologists interested in answering these kinds of questions about the learning and performance of mathematics, we must come to understand the structure of the subject matter itself, that is, we must understand something about mathematics as the mathematician views it. Without an understanding of the subject matter, we might pose psychologically interesting questions and yet never illuminate the psychology of mathematics as a discipline. On the other hand, if knowing mathematics were enough, then mathematicians would be adequate psychologists of mathematics. Their own introspections and their dialogues with colleagues would provide all we might want to know about how people learn and think about mathematics. For a true psychology of mathematics we need both the psychology and the subject matter. The mathematician establishes the subject matter, but the psychologist brings to the venture a knowledge of how people think in general and, even more important, of how to *study* how people think. It is this dual knowledge—knowledge of the structure of mathematics and knowledge of how people think, reason, and use their intellectual capabilities—that furnishes the ingredients for a psychology of mathematics. It is the study of how subject matter and human thinking interact that defines this field.

For a number of reasons it seems important today to define the psychology of mathematics as a special field of instructional research. For decades subject matter learning has not been a focus of psychological experimentation and theorizing. Instead psychology has concerned itself mainly with the universals of learning, thinking, and development, in the belief that general principles would automatically provide explanations for people's behavior in specific situations. Much of what we call the experimental method in psychology reflects psychologists' attempts to unearth these universal facts about how people learn, think, and develop. For example, for many years psychologists wished to study people's ability to remember things, but they wanted this study uncomplicated by memory for particular kinds of information and uncomplicated by existing knowledge that an individual might bring into a laboratory experiment. So psychologists looked for experimental tasks that would reveal as "pure" a memory as possible. This meant they sought tasks with little meaningfulness in terms of a real-world context for performance. And so it came about that the learning of nonsense words, strings of numbers, and so on emerged as important tasks in the experimental psychology laboratory.

Much important knowledge has been gained by using these isolated laboratory tasks, and it is reasonable to expect that much will continue to be so learned. We know a great deal more today about the general characteristics of human learning—the conditions under which people learn, how they remember things, and how current learning interacts with memory of the past—than we did before these experiments were begun. But we do not know much about how to put this information back into its natural context—the performance of real-life tasks—so that it tells us as much as possible about how particular subjects, such as mathematics, are learned.

Experimental psychology has not always shied away from specific subject areas. There was a period relatively early in the history of American experimental psychology when, in addition to laboratory studies, psychologists conducted direct research on the learning of school subjects such as reading and mathematics. Early in this century, the work of such scholars as Edward L. Thorndike represented the beginnings of an educational psychology grounded in subject-matter learning. In the 1930s, however, experimental and educational psychology began to draw apart, with educational psychology often leaving the "parent" psychology departments and moving into the universities' growing schools of education. The effect was to isolate educational psychology from the newest developments in academic psychology and, simultaneously, to discourage "real" or "tough-minded" psychologists from attending to pedagogical applications of their research.

As a result, the study of subject-matter learning by psychologists went out of style for many years. During that long period, psychologists interested in education assumed that their job was to present as clearly as possible the major facts and theories of general experimental psychology. They assumed that educators would then apply those universal, scientifically derived facts directly to instruction. Educational psychologists contributed to the fundamental knowledge that made up the psychology of human learning and development, but they paid little attention to the specific learning tasks of mathematics, reading, or social studies.

During the 1950s, however, conditions began to change. Behavioral psychologists became interested once again in instructional questions and some, notably B. F. Skinner and his students, began systematically to apply the principles of behavior analysis and reinforcement theory to education. They developed programmed learning techniques and they designed classroom environments that attempted to reinforce learning from both the social and intellectual standpoint. At the same time, experimental psychologists began to expand their focus of study to include nonobservable behaviors such as reasoning, thinking, and problem solving, and a new field of experimental psychology—cognitive psychology—was born. One arena in which these complex intellectual processes could be seen to operate was in the performance and learning of specific tasks such as those encountered in school.

Today, then, alongside experiments that use very simple tasks to isolate certain pure skills from the world outside the laboratory, psychologists are studying what people do when faced with the kinds of complex tasks they routinely perform at school, at work, and at play. Some of these tasks are mathematical. As a result, it is again possible to point to a psychology of subject matter, specifically, the psychology of mathematics. It is only an emerging psychology, but its outlines are clear enough today to permit us to examine work that has been done and that needs to be done to guide the development of scientifically based instruction in mathematics.

PSYCHOLOGY AND THE TASKS OF INSTRUCTION

How might instruction be affected by a psychology of mathematics? In the past, teachers and instructional designers have had to integrate subject matter and psychological knowledge on their own. They have been given information on how humans think, how they develop, and so on— in other words, the universals that the experimental method offers with respect to the nature of human cognitive processes. Most teachers have also had at least a little training in the discipline of mathematics. But there has been no organized body of knowledge available to help in the task of devising a system of instruction from these two apparently separate and unrelated bodies of knowledge. Today, a psychology of mathematics that deals directly and explicitly with the interaction between the structure of the subject matter and the nature of human thinking can provide a basis for developing theory and instructional practice in this field. To the extent that psychologists can successfully describe what people do when they perform mathematical tasks and how they learn to think mathematically, to that extent will the psychology of mathematics be useful in teaching. There is still much to be done, of course, but the ground has been broken and it is possible to envisage not only a well-developed psychology of mathematics but also a science of instruction based on that psychology. This book attempts to convey both what is now known and what might someday be included in a psychologically based science of mathematics instruction.

The organization of this book reflects the lack of consensus among mathematicians and mathematics educators about what ought to be the proper emphasis in mathematics instruction. Many definitions of the subject matter of mathematics have been put forth, each leading to a somewhat different set of psychological questions and research strategies. For many years mathematics in the schools really meant arithmetic or calculation, with only a small amount of formal algebra and geometry offered toward the end of secondary school. In recent decades, the trend has been to introduce more and more sophisticated mathematics into the elementary and secondary school curriculum and, at the same time, to place increasing emphasis on fundamental concepts of mathematics—ideas such as sets, operations, functions—at younger and younger ages. Computational and conceptual approaches to mathematics instruction have existed in a kind of uneasy balance as mathematicians and educators have pressed for increased conceptual understanding of mathematics. Computation still forms a major part of the elementary school curriculum, but much attention has also been paid to ways of teaching fundamental mathematical concepts, even to very young children. To add further complexity to the picture, we can distinguish an approach that is concerned less with the content of mathematics than with mathematics as a form of thinking and reasoning. By viewing mathematics as thinking, one is led to focus on problem solving and discovery not just as a means of teaching mathematical concepts but as a major goal of mathematics education.

The chapters of this book take up the first two major definitions of mathematics: computational and conceptual. Chapters 2 through 4 treat mathematics as a set of computational skills corresponding to what we know as arithmetic. Chapters 5 through 8 treat mathematics as a set of integrated concepts, rules, and procedures that make up a structured discipline. The third definition, mathematics as problem solving and reasoning, is interwoven through all the chapters but is highlighted particularly in Chapters 6 and 8.

In each chapter, we present a major theoretical approach to the study of mathematical thinking, learning, and performance; examine relevant psychological research; and consider the implications of that research for instructional practice. We draw upon research from several different "schools" of psychology. Thus one will find represented behaviorist and associationist psychology, Piagetian psychology, gestalt psychology, and cognitive psychology. Because most current work has its roots in past psychological research, we present important points of view from the past, as well as those of more recent investigators. In each case, we try to relate historical material to current information-processing models of human memory and performance that are described in the course of the book.

Our presentation assumes little prior knowledge of psychological research. We therefore devote a portion of each chapter to research methodologies and general assumptions related to the particular psychological approach being considered. Our attempt throughout is to stay close to the task of developing and explicating a special kind of psychology that is directly concerned with the subject matter of mathematics. For readers who possess more psychological than mathematical sophistication, this book is intended as an introduction to some important questions that define mathematics as a topic for research. Readers with more knowledge of mathematics should find it possible to learn some psychology in the context of a subject they already know. In either case, we intend to convey a sense of how psychology can shed light on mathematical thinking and learning, how mathematics can pose questions for psychology, and how the resultant "psychology of mathematics" may guide instructional design and practice.

MATHEMATICS AS COMPUTATION

One way to define mathematics is as a body of computational rules and procedures. To the layman such a definition seems quite natural and, in fact, this definition pervades most of mathematics instruction in our elementary schools. If one asks a 10-year-old girl, "What are you learning in math these days?", she is far more likely to mention long division, multiplication tables, or addition of fractions than she is to mention sets, commutativity, inequalities, and functions. Elementary school mathematics is dominated by computation, and computational proficiency remains a major goal for instruction, despite efforts to reform the mathematics curriculum in the 1960s. The aim of the reform movement was to introduce into mathematics instruction as early as possible certain basic concepts of the discipline, such as the properties of the real-number system, simple geometric constructs, logic, and set theory (see Chapter 5). The reformers thought that if sufficient time and thought were devoted to teaching children the underlying constructs of mathematics, computational skill would follow along rather nicely. They assumed that computational proficiency could be developed without devoting a lot of time to computation in a direct way.

Curriculum changes notwithstanding, there is no doubt that computation still constitutes the bulk of children's experiences in elementary school mathematics. We expect children to know how to perform complex computations and to do them quickly and accurately. We also aim, but less successfully, at making them capable of applying these computational skills in problem solving.

Let us look briefly at the computational skills themselves. By computation, we mean addition, subtraction, multiplication, and division. We also mean the use of percents, fractions, and certain other basic, everyday-life kinds of skills. We mean, in short, what has traditionally been called *arithmetic*. In thinking about arithmetic or computational skills, it is useful to distinguish between simple associations, sometimes called the *number facts,* and complex procedures, called *algorithms,* in which a fixed sequence of operations has to be performed. The simple associations or number facts are the tables—the addition, subtraction, multiplication, and sometimes division tables—that most children are expected to memorize in elementary school. The complex procedures or algorithms are the series of separate steps required, for example, in performing long division or subtracting fractions with different denominators. Individuals executing such procedures must know what steps to perform, must perform them in the proper order, and must recall needed number facts accurately. When children learn to use number facts and algorithms in problem solving, they typically do so in the context of verbally stated problems, known as *word problems* or *story problems.* Word problems require children to interpret the words of the problem, set up an equivalent mathematical calculation, and then apply relevant procedures.

In the next three chapters, we consider what psychology offers to the instructor whose goal is to develop these kinds of computational skills in children. We look first at the historical and current uses of drill and practice as a means of developing speed and accuracy in computation. We then show how computational tasks have been analyzed into component skills using the method of hierarchy generation, and we suggest how informal task analysis may be used by teachers to plan the organization of instruction. The final chapter in Part I introduces a variety of methods for studying the complex mental processing that goes on while children are carrying out computational procedures.

2 The Psychology of Drill and Practice

When we think back on our own school days, to the hours we spent on arithmetic, many of us remember laboring over pages of problems. Often these were pages of identical calculations, where only the numbers were varied. Or we worked with flash cards until we could shout out the answers immediately and with no mistakes. This kind of work was called "drill and practice." It was supposed to help us achieve perfect mastery of basic addition, subtraction, multiplication, and division. It was to ensure that we would forever remember how to perform the arithmetic operations we had been taught.

Drill and practice has a place of long standing in the history of mathematics teaching, especially in arithmetic. At one time it was the major means of instruction. Today it is still part of the mathematics curriculum, although usually accompanied by concrete experiences or explanations of underlying mathematical principles. Most everyone accepts some form of practice as necessary. The reason, according to educators and lay people alike, is that "practice makes perfect." Along with drill and practice come increases in speed and accuracy, which are two widely accepted criteria of computational proficiency. If children can execute calculations speedily and accurately, most people are satisfied that they "know" their computational skills.

What do psychologists know about the role of drill in establishing and maintaining computational proficiency? This chapter explores the historical and theoretical bases for including drill and practice in the mathematics curriculum. We begin with a look at a psychological theory—associationism—that provides one theoretical justification for the use of drill exercises. We have chosen to focus on E. L. Thorndike, who is, in a sense, the "founding father" of the psychology of mathematics instruction. As a psychologist Thorndike was firmly rooted in a tradition of laboratory experimentalism; but he was also strongly committed to the task of translating laboratory findings into guidelines for classroom instruction. We also present the arguments advanced against drill methods by another psychologist, William Brownell. Because of its prominence in arithmetic teaching, practice has received a great deal of study, especially during the first half of this century, most of it geared to make drill better organized and more effective. We survey a line of research that attempted to determine the relative ease or difficulty of arithmetic problems—and to account for those differences—so that teachers could plan the proper amounts and sequences of practice. We describe a computer-assisted drill program as an example of one way psychologists have attempted to optimize amounts of practice and rates of progression through drill material. Finally, we present theory and research that indicate why it might be important to develop speed and accuracy in certain kinds of computations.

EDWARD L. THORNDIKE AND THE FORMATION OF BONDS

In 1922 a small book appeared, called *The Psychology of Arithmetic*. It was written by Edward L. Thorndike, a psychologist working at Teachers College of Columbia University, who helped to develop some early principles of stimulus–response learning psychology. Thorndike is perhaps best known in psychology for his statement of the *law of effect,* an early version of what we now call *principles of reinforcement.* He discovered this law, not in the context of a complex subject like mathematics but in the context of simple laboratory experiments with cats, dogs, monkeys, and chickens.

According to Thorndike, in any given situation an animal had a number of possible responses, and the action that would be performed depended on the strength of the "connection" or "bond" between the situation and the specific action. The experiment most frequently associated with this idea involved placing a cat in a wooden puzzle box that could be opened by tripping a latch. Naturally, the cat would object to being confined in such close quarters and would claw and scratch at the side of the box to get out. Eventually it would accidentally trip the latch, opening the door and escaping. Replaced in the box, the cat would again claw and scratch; but each time the experiment was repeated, the cat took less time to find its way out. Of all the clawing and scratching responses, only the one that opened the door was rewarded by the opportunity for escape. In Thorndike's conception, the cat was not "figuring out" how to open the box; rather the reward of escape was serving to strengthen the bonds between the experimental situation and the particular response that permitted escape. Hence Thorndike's formulation of the law of effect (Thorndike, 1913): "When a modifiable connection between a situation and a response is made and is accompanied or followed by a satisfying state of affairs, that connection's strength is increased: When made and accompanied or followed by an annoying state of affairs, its strength is decreased [p. 4]."

Though he experimented mostly with animals, Thorndike thought his learning principles should apply equally to humans. Along with many other psychologists of the time—called "connectionists" or "associationists"—Thorndike argued that all human behavior, thought as well as action, could be analyzed in terms of two simple constructs. When broken down into its most basic units, behavior would be found to consist of *stimuli*, or events external to the person, and *responses*, or things that people did in reaction to those external events. When a certain response was given to a certain stimulus and followed by a reward, then a bond, or association, began to be formed between the stimulus and the response. The more frequently a certain stimulus–response pair was rewarded, the stronger the bond. Thus the law of effect—a special case of the laws of association—suggested that practice followed by reward was an important way in which human learning took place.

Associations between stimuli and responses, bonds, and the law of effect—how could these principles, developed largely by observing animals perform the simplest of behaviors, be applied to something as complex as school learning? That was the question that Thorndike (1922) addressed in *The Psychology of Arithmetic*. The answer seemed straightforward because associationism held that all knowledge, even the most complex, was built of these simple connections. Learning thus consisted of establishing and strengthening the needed associations. "The aims of elementary education," Thorndike said, "when fully defined, will be found to be the production of changes in human nature represented by an almost countless list of connections or bonds whereby the pupil thinks or feels or acts in certain ways in response to the situations the school has organized and is influenced to think and feel and act similarly to similar situations when life outside of school confronts him with them [p. xi]."

Rather than simply announcing the laws of learning to teachers and educators, Thorndike set out to demonstrate how they could be applied to the problems of instruction. What teachers needed, he believed, was to find and make explicit the particular set of bonds that constituted arithmetic. Once well-organized lists of all these bonds could be drawn up, then rewarded practice would enable the law of effect to strengthen these bonds, and one could expect improved performance in arithmetic. Thorndike's book was an attempt to explain how the subject matter of arithmetic could be translated into psychologically formulated stimulus–response bonds.

Because children of elementary school age were not yet able to deduce the rules of arithmetic from examples and other rules, Thorndike reasoned, the task of instruction was to form carefully the necessary bonds and habits that would allow them to perform computations and solve problems. As a first step, one would have to select the bonds to be formed. Naturally, any carefully constructed

14 2. THE PSYCHOLOGY OF DRILL AND PRACTICE

arithmetic curriculum, with or without benefit of psychological analysis, would divide the subject matter up into broadly defined topics. For example, multiplication would be treated as a composite of abilities, such as: "knowledge of multiplication tables up to 9 × 9; ability to multiply two (or more) place numbers when carrying is not required and no zeros occur in the multiplicand; ability to multiply by 2, 3, . . . , 9, with carrying;" and so forth, up to the ability to multiply two-place decimals (as with United States money), with fractions, and with mixed numbers. What Thorndike, as a psychologist, proposed was to analyze these abilities further into a detailed set of mental habits or connections, each of which would become a candidate for formation and strengthening. Figure 2.1 shows an analysis of simple addition in columns, of which Thorndike (1922) says: "The majority of teachers probably treat this as a simple application of the knowledge of the additions to 9 + 9, plus understanding of 'carrying.' On the contrary there are at least seven processes or minor functions involved in two-place column addition, each of which is psychologically distinct and requires distinct educational treatment [p. 52]."

Once the proper bonds were selected, how could they be formed and strengthened? This was where drill and practice came in. Proper drill and practice, according to Thorndike, involved practicing bonds in a carefully programmed way so that important bonds were practiced often, and lesser bonds, less often. So-called "propaedeutic" bonds, used only to facilitate learning new concepts, would be practiced temporarily but later drop out from disuse. For example, to add four 5's in a column, a child might be taught a propaedeutic bond like counting 5, 10, 15, 20; however, because this was to be replaced later

---

Learning to keep one's place in the column as one adds.
Learning to keep in mind the result of each addition until the next number is added to it.
Learning to add a seen to a thought-of number.
Learning to neglect an empty space in the columns.
Learning to neglect 0s in the columns.
Learning the application of the combinations to higher decades may for the less gifted pupils involve as much time and labor as learning all the original addition tables. And even for the most gifted child the formation of the connections '8 and 7 = 15' probably never quite insures the presence of the connections '38 and 7 = 45' and '18 + 7 = 25.'
Learning to write the figure signifying units rather than the total sum of a column. In particular, learning to write 0 in the cases where the sum of the column is 10, 20, etc.
Learning to 'carry' also involves in itself at least two distinct processes, by whatever way it is taught.

FIG. 2.1 Thorndike's analysis of column addition into bonds. (From Thorndike, 1922.)

by the bond "four 5's are 20," it would receive only minimal practice. Bonds were recognized to have an effect on each other; hence Thorndike (1922) noted: "Every bond formed should be formed with due consideration of every other bond that has been or will be formed; every ability should be practiced in the most effective possible relations with other abilities [p. 140]." The reward that served to strengthen the practiced bonds was obtained when arithmetic problems were made interesting, fun, and close to practical applications. Thus, Thorndike was also concerned with the intrinsic meaningfulness of problems and their relevance to daily activities outside of school.

Some of Thorndike's bonds seem quite straightforward. It is easy for us to imagine that learning a bond like "2 + 2" (the stimulus) equals "4" (the response) would be enhanced by appropriate forms of drill. Arithmetic is full of these simple bonds. But not all arithmetic is so easy to translate into stimulus–response terms. As anyone knows who has tried to learn long division, some of it involves extremely long and complex operations. Thorndike (1922) explained these complex operations as "organized cooperating system(s) of bonds [p. 138]," groups of individual bonds that needed to be taught "teamwork" by being carefully sequenced and practiced, as described in the following quotation:

> As each new ability is acquired, then, we seek to have it take its place as an improvement of a thinking being, as a cooperative member of a total organization, as a soldier fighting together with others, as an element in an educated personality. Such an organization of bonds will not form itself any more than any one bond will create itself. If the elements of arithmetical ability are to act together as a total organized unified force, they must be made to act together in the course of learning. What we wish to have work together we must put together and give practice in teamwork [p. 139].

If, as Thorndike suggested, bonds were created by repeated pairing of stimuli and responses, then it seemed the teacher's job was merely to provide the proper amount of practice, in the proper order, on each class of problems. The teacher was to identify the bonds that made up the subject matter of interest and then put them in order, easier ones first, arranging them so that learning the easier ones would help in learning the harder ones that came later in the series. When that was accomplished, all that remained was to arrange for children to practice each of the kinds of bonds. Each class of bonds was to be practiced just enough so that errors could be avoided when advancing to the next harder class of bonds. The more often drill and practice could be presented in the context of interesting and practical problems, the stronger would be the connections. And since more complex problems were conceived as strings of bonds, it was important to drill well on each of the connections that would be needed for the harder problems.

Thorndike's books were rich with examples of the specific problems and drill sequences he recommended (an example appears in Fig. 2.2); but the rules for generating them were largely intuitive. Which bonds were easier? What was

16   2. THE PSYCHOLOGY OF DRILL AND PRACTICE

---

32. A Percentage Race

Each row of pupils is a team. The teacher gives out printed problems, or uses those on these pages, or writes problems on the blackboard. All start together and write the missing numbers or answers as quickly as they can without making a mistake. At the end of 10 minutes all stop. The pupils interchange papers, mark with a cross each wrong result, and count the number of correct results. A pupil's score is the number of right answers with 2 off for each one wrong. The row with the highest average wins. Each pupil who makes any mistakes corrects them at home or during the study hour. Practice with this and the following page until you can make a good score.

1. 15% of $1.50 = ...
2. 12% of $2.15 = ...
3. 20% of 80¢ = ...
4. 4% of $300 = ...
5. 3 1/2% of $16 = ...
6. 1/2% of $400 = ...
7. 105% of $90 = ...
8. $14 = ... % of $20.
9. 39 = ... % of 78.
10. 56 = ... % of 60.
11. 16 = ... % of 25.
12. 5 = ... % of 7.
13. 8 = ... % of 9.
14. 16 = 20% of ...
15. $30 = 4% of $...
16. $75 = 5% of $...
17. $5 = 10% of $...
18. $12 = 6% of $...
19. 6% of $2000 = ...
20. 4 1/4% of $24.50 = ...
21. 1 1/2% of $6000 = ...
22. 76 = ... % of 380.
23. 22% of 25 mi. = ...
24. 4 = ... % of 11.
25. 1/2% of 600 = ...
26. 3% of 16 mi. = ...
27. 15% of 8 hr. = ...
28. $25 = ... % of $130.
29. 83 1/3 = ... % of $40.
30. 15 = 75% of ...
31. 2 1/2% of $450 = ...
32. 3/4% of $760 = ...
33. 45 = ... % of 80.
34. 72 = ... % of 80.
35. 140 = ... % of 215.
36. 122% of $64.50 = ...
37. 18 = ... % of 40.
38. 1/8% of $1000 = ...
39. 21 = ... % of 40.
40. 21 = ... % of 15.

FIG. 2.2   A sample drill lesson designed by Thorndike. Note the use of a "team race" approach, one way of strengthening bonds through reward. For this particular race, children were to complete 100 problems in 10 minutes. (From Thorndike, 1924.)

"enough" practice? What was the best way to organize practice on different kinds of bonds? These questions were not systematically addressed by Thorndike, but they generated a great deal of research on the psychology of arithmetic. As we see in a later section, this research continues to influence instructional practice even today. Thorndike thus took a big step in the direction of bringing psychological theory to bear on instruction. His contribution to the psychology of mathematics was to focus attention on the *content* of learning and to do so in the context of a specific subject matter. Thorndike's analysis of arithmetic in terms of bonds to be strengthened in what children needed to learn in mathematics and how they could learn it.

DRILL VERSUS MEANINGFUL INSTRUCTION

Thorndike's theory of learning, and the drill method of instruction that it seemed to promote, was not without is detractors, even within associationist psychology. Although Thorndike stressed that drill problems should be made interesting and should be verified with concrete objects, the thrust of his influence was to sanction drill as the major method of arithmetic instruction. Early on, a strong voice was heard opposing the bond theory—that of William Brownell, among others. Brownell objected to the drill method, which he saw as a direct extension of the bond theory, on several grounds.

First, it took no account of qualitative differences in the computations of children and adults. If an adult could compute a grocery bill by directly recalling a few addition facts from memory, did this mean a child should be taught to do it exactly the same way? Brownell thought not. When he interviewed third graders in drill programs (Brownell & Chazal, 1935/1958), he found them using a variety of procedures other than direct recall to do their addition and subtraction problems. They counted on their fingers; they solved from known combinations (e.g., I know that $4 + 4$ is 8, so $4 + 5$ is 9); or they gave immediate answers, but incorrect ones, which indicated they were simply guessing. The children were using these methods even after having been drilled for 2 years in the number combinations. Brownell interpreted this to mean that drill simply made them faster and better at the "immature" procedures they had discovered for themselves, not at the kind of direct recall that adults possess.

Second, the drill method implied a distorted view of the goal of learning. For Brownell and others of his persuasion (e.g., Wheeler, 1939) the criterion of arithmetic skill was the ability to think quantitatively, not to respond with 100% accuracy to a given list of arithmetic problems. Said Brownell (1928):

> The child who can promptly give the answer 12 to $7 + 5$ has by no means demonstrated that he knows the combination. He does not "know" the combination until he understands something of the reason why 7 and 5 is 12; until he can demonstrate to himself and to others that 7 and 5 is 12; until he is so thoroughly convinced that 12 is the right answer for $7 + 5$ that he can give it as the answer with assurance of its correctness; and until he can use the combination in an intelligent manner—in a word, until the combination possesses meaning for him [p. 198].

To ensure this sort of meaningfulness through instruction required some attention to the mathematical principles and patterns underlying computations. "If one is to be successful in quantitative thinking," Brownell said, "one needs a fund of meanings, not a myriad of 'automatic responses.'... Drill does not develop meanings. Repetition does not lead to understandings [Brownell, 1935, p. 10]." For example, asking children to recite "2 + 2 = 4" over and over again did not guarantee they could understand an important underlying number concept: The symbol "2" refers to the attribute of "twoness" that characterizes a set comprising that many objects; "2 + 2" represents the operation of combining two sets to create a set having the attribute of "fourness." Without plenty of experience combining and taking apart sets of concrete objects, a child reciting number combinations perfectly was just making "correct noises," Brownell thought. Instruction that stressed concepts and relationships was seen as the way to ensure skilled quantitative thinking. Though perhaps more roundabout than drill methods in achieving speed and accuracy, the meaningful method would achieve something more important. It would "help pupils organize and unify their knowledge of number, to develop facility in dealing with numbers, and to understand the principles of number combination [Brownell, 1928, p. 211]." Without meaningful instruction to point out the interrelationships, drill would encourage students to view mathematics as a "mass of unrelated items and independent facts."

Given the proper understanding of mathematical concepts and procedures, Brownell went on to say, students would be better able to apply their knowledge in novel situations. There was research to back up this assertion. For example, McConnell (1934/1958) compared a rigorous drill method (sheer repetition of abstract symbols) to a meaningful method of instruction. In the latter, the number facts were presented in conjunction with pictures or objects and then practiced with opportunities to verify answers. He found drill to be "the more forthright means of attaining automatic and immediate responses to the number facts," but on measures of transfer to untaught combinations, the meaningful approach gave significantly better results.

Later, Swenson (1949/1958) compared three methods for teaching the addition combinations to second graders. The *drill* method featured interesting, varied drill exercises; avoided excessive rehearsal of errors; and provided for extra practice on difficult combinations. The *generalization* method encouraged children to apply their learning to new problems at every phase and allowed them to continue counting and solving methods until they could switch comfortably to direct recall. The *drill-plus* method combined drill with some meaningful instruction: Number combinations were verified using counting and concrete manipulations; facts were arranged into groups that yielded the same answer, so as to highlight mathematical patterns. The generalization method proved most effective in promoting both learning of the material and transfer to new material, with drill-plus somewhat less effective, and straight drill least effective.

Presumably, at some point even children taught by "meaningful" methods had to practice number combinations and computational procedures so they could recall them accurately and quickly. Although Brownell declined to elaborate on the form, content, and timing of practice, focusing instead on the need for reform in arithmetic instruction, there did seem to be a place in his instructional scheme for some sort of practice. In concluding an extensive study of children's number concepts, Brownell (1928) hypothesized that computational proficiency developed in three overlapping phases: First, the child learned a procedure for executing a calculation—any procedure, be it counting on one's fingers, solving from known facts, or direct recall. There followed a period of increasing accuracy in the execution of the procedure, later accompanied by a rapid increase in the speed of calculation. Any switch to a new procedure—say, from addition by counting to addition by solving from known combinations—would be attended by an initial decline in accuracy and speed until the new procedure became familiar. This suggested one might introduce drill for accuracy and speed, but only following a period of familiarization with the computational process. Even so, this drill would have to be "meaningful habituation" rather than simple repetition. Practice, according to Brownell, would only be worthwhile if it included exercises to increase understanding.

Where Brownell and Thorndike differed was in their definitions of what should be learned. To Thorndike, mathematical learning consisted of a collection of bonds; to Brownell, it was an integrated set of principles and patterns. The two definitions in turn seemed to call for very different methods of teaching, either drill or meaningful instruction. Today most educators acknowledge the need for both types of learning experiences, but how they should be integrated is still not clear. In the meantime, a number of psychologists have elaborated the notion of meaningfulness in learning. In Chapters 5 through 7 we present more systematic theoretical justifications for meaningful instruction than those Brownell was able to offer, and we give some examples of what such instruction might entail. There we also consider a further argument against strict drill, namely, that it habituates children to an unthinking, inflexible mode of response, whereas mathematical thinking often demands flexibility and creativity.

WHAT MAKES ARITHMETIC PROBLEMS EASY OR HARD?

While drill methods came under increasing attack from psychologists of Brownell's persuasion, other psychologists applied their energies to the pursuit of better, more effective forms of drill. Thorndike and others examined the textbooks of the day and found them varying widely in the amount of practice given on the different number combinations. This spurred a period of research on the relative difficulty of arithmetic problems. The object was to be able to provide the proper amount of practice: less practice on easier problems, more practice on harder ones.

Knight and Behrens (1928), for example, monitored the behavior of 40 second-grade students as they learned and practiced the 100 addition and 100 subtraction combinations. These were the number combinations having a sum of less than 20, such as 1 + 2, 6 + 5, or 19 − 7. Knight and Behrens kept extensive records on the numbers of errors made, the numbers of exposures needed to master each combination, the amount of time needed to solve each combination on each occasion, and the amount of practice necessary to maintain proficiency. They considered children to have mastered the combinations as soon as they could reliably do them all with close to 100% accuracy, with no apparent hesitation, and with the same accuracy and speed after 3 weeks' elapsed time. The product of this study was a complete list of all the addition and subtraction combinations arranged in their order of difficulty—translatable into the amount of practice a teacher should require on each one. In subtraction, for example, they found that the hardest combination, 15 − 6, took children an average of 26 trials to master, at about 7.4 seconds per problem, with an average of 20 errors. A sample of their addition ranking is shown in Fig. 2.3. An interesting sidelight is that one could not assume that the reversed forms of number combinations were of comparable difficulty; in fact 9 + 5 ranked nineteenth, whereas 5 + 9 was the hardest of all; and 3 + 0 ranked twentieth, whereas 0 + 3 was the easiest of all.

To Brownell, data like these furnished grounds for criticizing the bond theory. If every single bond had to be individually taught—not only 5 + 9 but also 9 + 5 and later 9 + 15 and 9 + 25—then the task of instruction was absurdly immense. In the context of Knight and Behrens' drill program, children apparently did not learn, or did not know how to apply, the mathematical principle of commutativity. Later, we present evidence that, with the proper instruction, even preschoolers can demonstrate an understanding of commutativity.

Other researchers (e.g., Clapp, 1924; Wheeler, 1939) developed their own ordered lists of number combinations, and there were arguments over which list was more valid. The results of such studies were used to organize drill and practice in schools. Easier addition and subtraction combinations were introduced earlier—without necessarily giving attention to the logical relationships among combinations—and textbooks and exercise pages devoted greater or lesser amounts of space to each combination according to its ranking on someone's list. The same was true for the multiplication combinations (Norem & Knight, 1930).

Studies that simply ranked problems did not tell much about *why* the problems were easier or harder. An analysis of the patterns of ease and difficulty points to some possible explanations, however. For one thing, the combinations involving sums over 10 clustered around the harder end of virtually every list of addition problems. For another, subtraction problems took longer to solve and involved more errors than addition problems. Wheeler (1939) was able to show that problems having a common addend (e.g., all problems in which 7 is added to another number) were of approximately equal difficulty and that difficulty increased with the size of the addend. These findings suggest differences in the

What Makes Arithmetic Problems Easy or Hard? 21

Addition Combinations	Difficulty Rankings		
	Knight-Behrens (1928)	Clapp (1924)	Wheeler (1939)
5 + 9	100	90	82
7 + 9	99	99	100
8 + 7	98	92	97.5
5 + 8	97	98	97.5
8 + 9	96	89	99
9 + 7	95	97	94
7 + 8	94	93	91
8 + 5	93	100	84
4 + 9	92	85	78
6 + 8	91	96	96
.	.	.	.
.	.	.	.
.	.	.	.
9 + 2	55	9	64
3 + 8	54	70	66
2 + 5	53	23	43
2 + 7	52	63	45
8 + 2	51	34	48
5 + 4	50	12	53
2 + 9	49	24	59.5
2 + 4	48	40	50.5
4 + 0	47	18	22
7 + 7	46	15	46.5
.	.	.	.
.	.	.	.
2 + 2	10	3	37.5
6 + 0	9	36	13
0 + 7	8	50	17
0 + 4	7	43	13
0 + 1	6	49	2.5
2 + 0	5	14	9
9 + 0	4	75	6
1 + 0	3	59	9
1 + 1	2	10	9
0 + 3	1	54	32

FIG. 2.3 A sample of the difficulty rankings on the 100 addition combinations. Although used to specify the proper amount of practice on each combination, difficulty rankings also indicated possible differences in mental solution processes.

amount of mental processing required to do larger sums as compared to smaller or to do subtraction problems as compared to addition. The fact that a sum like 12 + 7 consistently takes longer or yields more errors than 7 + 2 suggests children may be using different procedures to find the two answers. Brownell's interviews with children revealed some of these possible procedures, and today psychologists are analyzing children's solution procedures at even finer levels of detail. In

22   2. THE PSYCHOLOGY OF DRILL AND PRACTICE

Chapter 4 we examine several different counting procedures that children use when they cannot recall the combinations directly from memory. The various procedures take different amounts of time depending on how large the number(s) to be counted are. These differences in processing may account for patterns of difficulty such as Wheeler (1939) found.

Alongside studies ranking the various arithmetic combinations, there was another line of research that attempted to rank problems on the basis of various measures of problem complexity. This research involved story or word problems, which are often used to give practice in specific computations, in general problem-solving skills, or in applications to real-life tasks. A word problem is a computation requested in verbal form, for example:

> Bob buys two boxes of nails at $1.00 each and 3 gallons of paint at $4.00 a gallon. If he had $20 to start with, how much does he have left to spend on Mary when he takes her to the movies this Saturday night?

Such a problem exercises multiplication (2×1, 3×4), addition ($2 + 12$), and subtraction ($20 - 14$) skills. It also gives practice in translating stories about real-life situations into mathematical problems.

The difficulty of word problems was shown to be affected by many factors. In a number of studies (Brownell & Stretch, 1931; Hydle & Clapp, 1927; Kramer, 1933), researchers presented children with word problems requiring identical operations on identical numbers but varying in wording and specific problem contexts. Some of the variables that seemed to contribute to problem difficulty, as determined by the number of errors children made, were the familiarity of the situations described in the problems; the arrangement of problems in a series; the number of unfamiliar objects and nonessential elements; whether the story problems were intrinsically interesting; how difficult the vocabulary was; whether the problems were presented as declarative statements or as questions; and so forth. On the basis of their research, Hydle and Clapp (1927) argued that word problems should involve familiar situations that children could easily visualize (e.g., a circus as opposed to an African diamond mine) as a first step in problem solving. Brownell and Stretch (1931) countered with evidence that situations did not need to be familiar, provided the other complicating variables were kept at a minimum level. In fact, they argued, some exposure to unfamiliar situations was desirable so that children would learn to appreciate the wide applicability of number operations. Although these studies were inconclusive, they did point to factors that teachers should be aware of in preparing or assigning word problems for student practice.

More recent studies have continued and systematized this line of effort, taking advantage of the data gathering capacities of computer-assisted instructional programs. Loftus and Suppes (1972), for example, predicted problem difficulty on the basis of what they called "structural variables," or characteristics of individual problems that seemed to contribute to their complexity. Each structural variable was quantified, and problems were assigned a difficulty rating accordingly. Sixth-grade students then worked on 100 word problems at a computer terminal that recorded all their solution attempts, correct and incorrect. In subsequent statistical analyses, certain of these variables were found to be particularly influential in determining story problem difficulty, as measured by the probability of a student's getting the correct answer: (1) the number of different arithmetic *operations* needed to arrive at the solution; (2) the *sequence* variable, or whether the problem was solvable by the same operations, in the same order, as the previous problem; (3) the *length* of the problem, or number of words in the problem statement; (4) the *depth* variable, or the grammatical complexity of the wording; and (5) whether or not a *conversion* of units of measurement was required. The effects of other hypothesized structural variables—required number of steps to solution, presence of verbal cues such as "and," "left," and "each" (signaling addition, subtraction, and multiplication or division, respectively), and the order in which information was presented within a problem—were not significant.

By the criterion of structural variables, the sample word problem about Bob and Mary should be judged fairly difficult. It requires three different operations (multiplication, addition, and subtraction), is 45 words long, contains embedded phrases as well as irrelevant information (Mary has nothing to do with the computational aspect of Bob's problem). However, it might be perfectly appropriate as a practice problem for a fifth or sixth grader. Contrast our statement of the problem with the way it might appear in some of the "pared-down" word problems in recent primary textbooks:

> Bob buys: 4 boxes of nails, $1 each
> 3 gallons of paint, $4 each
> Bob had $20. How much is left?

The same number of operations is required, but the number of words is considerably reduced and the grammar less complicated. The problem should be easier to solve and, therefore, more appropriate for younger or less skilled students.

The point of identifying these structural variables was to be able, eventually, "to formulate a clear set of rules or a formula for generating sets of arithmetic problems of a specified difficulty level. Curriculum developers would then be in a better position to control difficulty level when preparing instructional materials [Jerman & Rees, 1972]." In other words, the object was to design better practice sequences. At the same time, this research provided clues to the mental processing involved in certain kinds of problem solving. Knowing which variables affected problem difficulty suggested the kind as well as the number of mental steps needed for problem solution. But no theories specifying the steps involved in processing word problems were actually developed in the course of this research. In later chapters, we consider how detailed study of the steps by which story problems and other computations are solved has helped build psychological

OPTIMIZING THE EFFECTIVENESS OF PRACTICE

The relative difficulty of arithmetic problems is only one factor, albeit an important one, in organizing drill and practice. In addition to deciding which problems should be practiced first and which ones later, there are decisions to be made about when to introduce and terminate practice, when to switch to a new level of difficulty, what kind of practice to give, and the like. These questions accept the assumption that practice is important to becoming proficient in computation and they focus on organizing computational practice to optimize its effectiveness.

Psychologists have long been interested in optimizing practice on all kinds of skills, not just arithmetic. It is generally established, for example, that *spaced* practice is more effective than *massed* practice for most skills; that is, practice sessions of moderate duration spaced out over several days produce better learning than the same amount of practice concentrated into one long session. The superiority of spaced over massed practice for arithmetic in particular has been demonstrated experimentally by Buswell (1930) and Repp (1930, 1935). Repp was concerned with the form drill should take if one wanted to maintain children's proficiency in basic arithmetic operations on whole numbers and fractions. Should students be given extended practice involving only similar types of problems ("isolated" drill) or should different types of problems be interspersed with one another ("mixed drill")? Most textbooks of the day employed only isolated drill, but Repp felt there was little experimental evidence to support that method. He designed a study to explore the relative merits of isolated and mixed drill.

Repp tested 538 12-year-olds and created two groups matched on three levels of ability (low–average–high). One group of children practiced on blocks of similar problems, each week concentrating on one particular computational skill. This was the isolated-drill group. The mixed-drill group, on the other hand, practiced a variety of computational skills each week but worked only one or two problems of each type. Examples of the two forms of drill are shown in Fig. 2.4. Practice sessions took place once a week for both groups and lasted 20 minutes. After 26 weeks, both groups had received the same amount of practice on each type of problem, the only difference being the organization of practice. However, although both groups had gained in speed and accuracy, the mixed-drill group showed a 23% greater gain than the isolated-drill group. Mixed drill yielded greater speed and accuracy at all ability levels—low, average, and high alike—with the greatest gain (54%) appearing in the low-achievement group.

This experiment seemed to indicate that in general it was better to practice a skill in small amounts and frequently than to practice in great amounts less often.

The Isolated Type of Drill Organization

1. 2/9 + 5/9 =
2. 1/2 + 1/2 =
3. 5/7 + 2/9 =
4. 2/3 + 3/8 =
5. 9/10 + 1/5 =
6. 5/9 + 3/7 =
7. 8/9 + 1/6 =
8. 4/5 + 1/2 =
9. 2/3 + 5/6 =
10. 3/5 + 7/8 =
11. 3/5 + 2/7 =
12. 8/9 + 1/9 =
13. 7/8 + 2/3 =
14. 3/4 + 5/6 =
15. 3/5 + 1/2 =
16. 8/9 + 7/8 =
17. 1/2 + 1/4 + 1/3 =
18. 3/4 + 5/8 + 1/2 =
19. 5/7 + 2/9 + 3/5 =
20. 1/8 + 1/7 + 1/4 =

The Mixed Type of Drill Organization

1. 14) 8599
2. 6 1/2 − 1/3
3. 27) 538
4. 8963 × 38
5. 1/3 × 12 × 5 =
6. Subtract: 9 1/2 from 10 4/5 =
7. 80856 − 77184
8. 976965 9766 85 27234378
9. Multiply: 438 by 577
10. 295 × 389
11. 4 2/3 + 7 1/2 =
12. 3/10 ÷ 2/9 =
13. 84 31 99 46 77 665 3477 89 68
14. 3/4 × 2 × 1/3 =
15. 65) 1823
16. Add: 998 239 234 910 629
17. 9912 − 7383
18. Subtract 4 bu. 2 pk. 1 qt. from 8 bushels. Answer.................
19. Add: 94 1/2, 2/3, 26, 10 3/4
20. Multiply: 4209 × 63

FIG. 2.4 Examples of isolated and mixed drills. (From Repp, 1935.)

Nevertheless, after further study, Repp (1935) acknowledged that isolated drill seemed to play an important role in firmly establishing new learning. He suggested using isolated drills "in close proximity to the first teaching and learning of new facts, skills, and information," reserving mixed drill for maintenance purposes once the new skill was well learned. Isolated drill also seemed useful as an antidote to habitual wrong responses on particular computations, for example, in the case of a student who habitually reported 75 as the product of 9 × 8. Despite the many studies that existed on drill and practice at the time, including his own, Repp (1935) was able to conclude only that, "Generally speaking, the proper use of drill in arithmetic needs to be better understood. Correct drill construction and its use is still a fertile field for research [p. 200]."

Computer-Assisted Drill. The search for better organized, more effective drill programs continues into the present day. A new thrust has been added, however. With the availability of computer technology, some researchers have explored the possibility of tailoring drill programs quite precisely to the ability levels of individual children, and in so doing to optimize conditions for improving computational skill. Because of the computer's capability to interact directly with the learner and to store detailed data about the learner's performance, it has quickly been adopted as an experimental medium for individualizing drill and practice as well as other types of instruction. The resulting computer-assisted instructional (CAI) programs provide a unique opportunity to study the mathematical performances of children. As we see in later chapters, the computer has been an important tool in a variety of studies on human learning.

One series of drill-based programs that combined instruction with research on optimal approaches to teaching was designed by psychologists at Stanford University's Institute for Mathematical Studies in the Social Sciences. The Stanford arithmetic programs resulted from a project that used the computer as the basis for providing individual practice on the computational skills customarily taught in grades 1–6. (For a detailed description of these programs, see Suppes, Jerman, & Brian, 1968; Suppes & Morningstar, 1972.) Like the earlier noncomputerized experiments that studied mathematics learning in the context of instruction (e.g., Wheeler, 1939; Knight & Behrens, 1928), the Stanford work assumed that practice was essential if children were to become fluent and competent performers of computational tasks. And like the earlier studies, the Stanford programs assumed that drill should proceed roughly from easier to harder kinds of problems, include an element of speed demand, and strive for a high and lasting degree of accuracy.

There was nothing unusual about the content of the Stanford CAI programs. Like most standard school mathematics curricula, they contained practice in addition, subtraction, fractions, multiplication, long division, percents and ratios, and so on at various levels of difficulty appropriate to different age groups. However, unlike conventional programs, the computer was able to give continuing feedback, so that children could tell immediately whether their answers were right or wrong. It also had a built-in decision-making apparatus that allowed for adjustments in problem presentation and difficulty level. This meant children never had to work on problems that were too easy or too hard for them. Thus the CAI program represented a conscious attempt to enhance children's motivation while avoiding practice in error.

Briefly, this is how the programs worked from the point of view of the child and the teacher. In or near each participating classroom, a computer terminal was made available to students. Each child was scheduled to spend 5 to 15 minutes at the terminal once a day. Drill was organized into "concept blocks," to be completed in 3 to 12 days, that supplemented similar content taught in a textbook series. Each concept block contained problems of roughly the same type (e.g., sums from 0 to 20, multiplication tables 2 and 3, or units of measure), although many were "mixed reviews." Each included a pretest, 5 days of drill, a posttest, and sets of review drills and review posttests. The teacher assigned concept blocks appropriate to the material the class was covering in its arithmetic textbook.

Since all children in a class worked on the same concept block, they were all practicing roughly the same arithmetic content. But within each concept block the computer was able to present problems of five levels of difficulty; based on an individual child's performance, it could make a decision about which level of difficulty would be appropriate for practice. Each child started out at a level determined by his or her pretest score. On succeeding days, the difficulty of problems was adjusted according to the child's practice score from the day before. Children scoring between 60 and 79% stayed at the same level the next day; children scoring 80% or above moved to the next hardest level; those scoring below 60% were given easier problems. Review drills were "chosen" by the computer program so that individual children had extra practice in content on which they had scored least well in the past.

Feedback was given on each problem, as shown in the sample drill in Fig. 2.5. If a child's answer was correct, the computer printed out a message that said so; if a child gave the wrong answer, the computer reported the error, let the child try again, and after a certain number of tries printed out the correct answer. Further, if a child took too long to solve a particular problem, the computer gave a "time is up" signal to encourage the child to work at a reasonably fast pace.

We suspect that Thorndike would have been pleased by the possibilities for well-organized drill that the computer-assisted programs represented. Precisely tailored drills minimized the problem of "stamping in" incorrect bonds by forestalling errors as much as possible. Bonds that were more difficult—as determined by error data from daily performance—received the additional practice they required. But instead of determining the difficulty of bonds according to a subjective standard or some class average, the computer interpreted the difficulty of bonds according to the individual practicing them. Implicitly, it acknowledged that certain bonds might be harder for some people than for others. It was also able to respond to individual differences in previous knowledge and general aptitude.

Did the Stanford drill programs work? Data gathered on the Stanford programs during the most intensive period of study, 1965 to 1968, were used to

```
PLEASE TYPE YOUR NAME
MIKE ODELL
         DRILL NUMBER 509013
    (42 + 63) / 7 = (42 / 7 ) + (63 / 7)
    48 - 38 = 38 - 48
WRONG
    48 - 38 = 38 - 4
WRONG, ANSWER IS 28
    48 - 38 = 38 - 28
    76 - (26 - 10) = (76 - 26) + 10
    4 X (7 + 13) = (4 X 7 ) + (4 X 13)
    (53 - 20) - 11 = 53 - (20 + 11)
    32 + (74 + 18) = ( 32 + 74) + 18
    51 X (36 X 12) = ( 51 X 36) X 12
    17 X (14 + 34) = (17 X 14) + (17 X 34)
    362 + 943 = 943 + 362
    (5 + 8) X 7 = ( 5 X 7) + ( 8 X 7)
    (90 / 10) / 3 = 90 / (10 X 3)
    (72 / 9) / 4 = 72 / ( 9 X 4)
    (54 + 18) / 6 = (54 / 6) +(18 /__)
TIME IS UP
    (54 + 18) / 6 = (54 / 6) + (18 / 6)
    60 - (10   18) = (60 - 19) + 12
    72 X (43 X 11) = (72 X 43) X 11
    (63 / 7) + (56 / 7) = ( 63 + 7 ) ( 7
WRONG
    (63 / 7) + (56 / 7) = ( 63 + 56) / 7
         END OF DRILL NUMBER 509013
    13 MAY 1966
    16 PROBLEMS
                    NUMBER    PERCENT
    CORRECT           13         81
    WRONG              2         12
    TIMEOUTS           1          6
    WRONG
    2
    16
    TIMEOUTS
    13
       222.7 SECONDS THIS DRILL
    CORRECT THIS CONCEPT - 81 PERCENT, CORRECT TO DATE - 59 PERCENT
    4 HOURS, 46 MINUTES, 59 SECONDS OVERALL
    GOODBYE MIKE.
```

FIG. 2.5 Sample printout from the Stanford computer-assisted drill programs. This represents one drill session completed by a fifth grader. (From Suppes, Jerman, & Brian, 1968. Copyright 1968 by Academic Press. Reprinted by permission.)

analyze the performance of almost 4000 children, grades 1-6, in selected sites in four different states. The Stanford Achievement Test (SAT) scores of children who had participated in the CAI program were compared with the scores of control groups who had simply received standard classroom instruction. On most of the sections of the SAT test, and in most grades, the CAI children improved significantly more than did the control children. This was true even for tests that stressed concepts and applications rather than strictly computational skills. And this effect was obtained with only 8 months of work, and only 5 to 8 minutes per day at the computer. But these results were obtained using experimental groups who received classroom instruction combined with supplementary drill on the computer, whereas the control groups had only regular classroom instruction and no extra drill. What if the control classes had received supplementary drill in the form of paper-and-pencil practice exercises?

Through an unforeseen occurrence, the Stanford study was able to suggest an answer to this question as well. One of the control schools found their pretest scores were so poor that they initiated their own paper-and-pencil drill sessions. Children who had this extra work for 8 months actually did better on the posttest than the CAI children. This finding agrees with other more recent, deliberately planned evaluations of CAI versus paper-and-pencil drill programs (Jacobson, 1975). However, these control children had to put in 25 minutes per day of extra work, compared to the 5-8 minute computer sessions, and the teacher had to spend extra time grading their papers. Nevertheless, these results do suggest that the practice itself, rather than the fact that it was occurring on a computer terminal, produced the improvement in performance.

The Stanford data show that well-planned drill and practice can increase accuracy in computations of many kinds for children of many ages. However, children who started out with better accuracy scores improved about as much as did the children who started out lower in accuracy. Thus, although the programs adapted to individual differences, they did *not* have the hoped-for effect of helping the initially less competent children "catch up" with the more competent ones. Instead, all children improved in most areas. What is more, some concepts were very hard to improve within the confines of the practice programs. This suggests that drill will not improve all aspects of arithmetic performance, not even drill that responds to individual differences.

It is noteworthy, too, that the Stanford programs rarely demonstrated accuracy greater than 90% within concept blocks (Suppes & Morningstar, 1972), a result consistent with other learning experiments conducted in the laboratory (Judd & Glaser, 1969). In a study on computer instruction in simple number combinations, Jacobson (1975) demonstrated that once accuracy reached a certain level, 90-95%, further practice could bring about no significant increase. This is quite far from the 995 to 997 out of 1000 mastery Thorndike thought could be achieved! These studies suggest it may never be possible to achieve perfect

2. THE PSYCHOLOGY OF DRILL AND PRACTICE

accuracy, at least for children. Teachers and instructional designers thus need to be sensitive to the point beyond which additional drill has no particular value.

DRILL AND THE DEVELOPMENT OF AUTOMATICITY

Much of the theory and research presented so far has concerned the role of drill in promoting speed and accuracy. We can assume that accuracy is important in computation, but one might well ask whether speed is a particularly important goal of instruction. As we have noted and as we develop further in the second half of this book, some psychologists and educators argue that instruction on even the simplest, most basic arithmetic skills should help children understand mathematical concepts rather than simply memorize facts and procedures. If understanding is established, they argue, children can reconstruct the forgotten items or even construct their own procedures for finding answers when memory fails. Where the criterion of mastery is grasp of ideas, speed of recall seems comparatively unimportant.

The counterargument holds that, on the contrary, it is very important for children to memorize certain facts and procedures to the point where they need not think about them but can do them rapidly and almost automatically in the course of computation. The function of drill, according to this view, is to develop automatic responding, indicated by a high rate of speed. Automatic responding has been talked about in psychology since before the days of Thorndike (e.g., Huey, 1908, on the role of automatic responses in reading). However, recent developments in psychological theory now make it possible to be more specific about what we mean by automatic responding—or *automaticity,* as it is often called.

A convenient framework for considering automaticity, and one that we call upon in the ensuing chapters, is the information-processing view of the human mind. In this view all human behavior is seen as the result of the mind acting upon (processing) data from the internal or external environment (information). Although there are many differences in detail, virtually all psychologists working within the information-processing framework hold a common assumption about the structure or "architecture" of the human mind. The assumption is that information is processed through a series of "memories," each capable of different kinds of storage and processing and each subject to different limitations. Together these memories constitute the information-processing "system."

Working from the outside in, information first enters the system through a *sensory intake register* (sometimes called a sensory buffer or iconic memory). This first memory in the system can receive visual, auditory, and tactile information directly from the environment, and it can take in a lot of information simultaneously. But it can hold that information for only a very short time—less than 1 second. If in that time the other components of the memory system fail to attend to the information in the sensory intake register, that information is lost. The component that does this "attending" is called *working memory* (or sometimes short-term memory or intermediate-term memory). This is where the actual thinking gets done, that is, where operations are performed on information. The third component of the system is *long-term memory*, where everything a person knows is stored.[1]

Within this general structure, working memory plays a crucial role. Only by being processed in working memory can information from the sensory part of the system enter a person's long-term memory store. And only when information is called out of the long-term store into working memory can the stored information be used in the course of thinking. Like the sensory intake register, working memory has a limited processing capability. Working memory is limited not by the length of time in which it can retain information but by the amount of information it can handle at any one moment. No one knows exactly how many "pieces" of information it takes to "fill up" working memory, although about seven pieces, give or take two, has been long proposed as the capacity of adults' working memories (Miller, 1956). If working memory is "full," then new information coming in from either the sensory system or long-term memory is accepted, but older information is lost. In general, the information that has been least recently attended to is the information that will be lost. Rehearsal (repeating the information to oneself from time to time) can make it possible to retain information in short-term memory. But rehearsal cannot increase the basic capacity; it cannot extend the number of memory "slots" there are to be filled. One way to extend working memory's processing capacity, however, is to organize small pieces of information into "chunks," so that each slot is in fact filled with more information. The larger the chunks, the more information working memory can handle. We explore these organizing or chunking processes later in Chapters 4 and 8.

Another way to extend the capacity of working memory is by developing automaticity of responding. The argument is as follows: To the extent that certain processes can be carried out automatically, without need for direct attention, more space becomes available in working memory for processes that *do* require attention. To relate automaticity to the domain of computation, we need to distinguish between two kinds of arithmetic tasks on which drill and practice is commonly given. On the one hand we have the so-called *number facts*, that is, the single-digit number combinations that form the basic building blocks of all computations. Number facts are of four kinds—addition, subtraction, multiplication, and division (e.g., $5 + 4 = 9, 8 - 3 = 5, 9 \times 8 = 72, 8 \div 2 = 10$). On the other hand, we have *algorithms*, or procedures of computation. These are the

---

[1] An alternative conception (Craik & Lockhart, 1972) plays down the separate character of these memory stores in favor of the notion of successively deeper "levels of processing," but all the fundamental memory operations must still be carried out.

sequences of operations that we perform, using the number facts, to arrive at solutions to more complex problems. For example, to add columns of digits that sum to more than 10, we perform a series of smaller operations:

To add: 20   We do the following:
        18   Add the ones column ($0 + 8 = 8$;
         4    $8 + 4 = 12$; $12 + 3 = 15$).
         3
Notate the 5; carry the 1.

Add the tens column (1 [carried] + $2 = 3$; $3 + 1 = 4$).

Notate the 4; answer 45.

This procedure is an algorithm. To take another example, long division involves estimating divisibility, multiplying, subtracting, bringing down the next number, dividing again, and so forth. This is the long-division algorithm.

Suppose Susan must execute an algorithm involving several steps, such as adding fractions of different denominators:

$$3\tfrac{1}{2} + 8\tfrac{1}{4} = ?$$

The actual sequence of steps is long, and an information-processing psychologist would say it requires holding a great deal of information in working memory, particularly if it must be done without paper and pencil:

$(3 \times 2) + 1 = 7$, over 2, change to fourths, equals 2×7, over 4, or 14 fourths; plus $(8 \times 4) + 1$, over 4, or 33 fourths; $14 + 33$ over 4, etc.

While Susan carries out this calculation, much of working memory will tend to be occupied with keeping her place in the long procedure and holding the numbers needed for several intermediate computational steps. It is easy to see in this case why it might be desirable for Susan to have an automatic command of the number facts. If space must be taken up in working memory by the computation of 8×4 or 2×7 or by an extended search for these facts in long term-memory, then place-keeping functions and memory for intermediate computations are likely to meet serious interference. What is more, procedural errors are likely to creep in unnoticed. On the other hand, if practice has rendered those number facts instantaneously retrievable from long-term memory, then working memory is permitted to function more efficiently.

The same holds for automatic access to memorized procedures or algorithms. If Susan has to reconstruct how to change fractions into their lowest common denominator each time this procedure is needed, then valuable space is again taken up in working memory by processes that might well have become automated through proper practice. Thus emerges the strong suggestion that at least certain basic computational skills—number facts and simple algorithms—need to be developed to the point of automaticity so they can avoid competing with higher-level problem-solving processes for limited space in working memory.

Direct evidence regarding the effect of automaticity on arithmetic calculation is not yet available. However, there exists analogous evidence in the psychological research on reading. It seems that automaticity of word recognition skill is associated with higher levels of reading comprehension (LaBerge & Samuels, 1974; Perfetti & Hogaboam, 1975). Perfetti and Lesgold (1979) encountered three types of word recognition skill in young readers, each of which suggested a different instructional treatment: (1) slow and inaccurate; (2) slow but accurate; and (3) fast and accurate. If we extrapolate to the domain of mathematics, we might find similar levels of skill among, say, learners of number facts. We could predict what kind of practice would benefit each type, as follows: Students demonstrating slow, inaccurate mastery of number facts would need practice that stressed accuracy and perhaps some help in understanding the reasons for particular procedures. Fast, accurate students, on the other hand, would need no isolated practice at all. The slow but accurate students would be candidates for exercises specifically designed to develop automaticity, for example, speeded practice. Of course, we are speaking hypothetically, based on studies in the domain of reading. But there is promise in this approach for new insights into the enhancement of mathematical skills.

It is clear from recent surveys that many people, even adults, perform very poorly in arithmetic calculation. Teachers seem to feel this is due to incomplete or inadequate mastery of number facts (Jacobson, 1976; Lankford, 1972), and psychological studies support this contention (Anaspaugh, cited in Buswell, 1927; Tait, Hartley, & Anderson, 1973). In other words, when children's algorithmic computations are examined, the correct processes have often been carried out in the proper order, but specific errors in recall of number facts produce errors in the final answers. However, in at least one study (Jacobson, 1975) that compared different types of drill, including CAI, accuracy of performance on number facts did not seem to predict performance on more complex algorithms. This suggested that errors in algorithmic computations could not be attributed solely to not knowing number facts. Jacobson sought to explain the discrepancy by the fact that isolated practice of simple number combinations presented unique learning and memory problems that do not exist when the combinations are used in actual problem-solving situations. In our view, another contributing factor could have been a lack of automaticity for number facts. Slow, accurate students, for example, might easily have passed accuracy tests on number facts but would still have failed in complex computations because their command of number facts was not automatic enough to avoid placing a heavy processing load on working memory.

A comment is in order concerning the way some of the recent innovative mathematics curricula incorporate drill and practice. The ideal espoused by the curriculum reform movement of the early 1960s (*Goals for School Mathematics,*

1963) was to avoid drill as much as possible by building it into the curriculum in a "spiral" fashion. Instruction in a new procedure or skill would be followed by a limited amount of drill on that skill by itself. Additional experience with that skill would later be provided, at various points in the curriculum, in the context of other types of problems of increasing complexity. The hope was to avoid the pitfalls of drill but to ensure enough practice so that competence in arithmetic skills was achieved.

The research we have presented suggests dangers that teachers should be aware of in the spiral approach. Students who fail to commit specific number facts or algorithms to memory the first few times they are presented may never "catch up" if the only further practice they receive is embedded in more complex problems. Imagine children trying to keep their minds on the complicated steps in finding square roots if they had to stop repeatedly and think what 6×7 was or how they were supposed to do "borrowing"! Drill and practice as a supplement to instruction—for the specific purpose of developing automaticity—may be of immense help from time to time for certain students, just to forestall such a pattern of failure and frustration.

One would like to be able to state a definitive set of rules for administering drill and practice. Unfortunately this would be premature. There are indications that drill does increase the speed and accuracy of responses, and we have pointed to one reason why speed might be particularly important in certain types of calculation skills. However, the research surveyed to this point leaves many questions unanswered. There are questions concerning what children learn: Do they learn bonds or patterns? Do these seemingly different types of knowledge demand different forms of practice? Could practice on a variety of related bonds eventually contribute to understanding larger patterns? There are questions concerning when practice should be introduced: Should it come before, during, or after more "meaningful" instruction in the conceptual underpinnings of computational procedures? Could repetition actually contribute to conceptual understanding? There are serious questions concerning what drill really accomplishes: Does drill teach children to do immature procedures faster, or does it push them to learn more efficient procedures? And how does the switch from one procedure to another come about? These issues lead us head on into the question that concerns us throughout this book: What goes on in children's heads when they perform mathematical tasks? We are interested in uncovering the step-by-step processes by which they arrive at solutions to mathematical problems. We want to know what it means for them to think mathematically.

When research provides a clearer picture of children's learning and thinking processes, the proper role of drill and practice should no longer be in doubt. But even so, a very practical question will remain—how to remove from the drill and practice its negative connotations. Should psychology and education manage to narrow the range of applicability of drill, to systematize guidelines for its organization, and to design maximally effective practice sequences, the situation will be already be much improved. But all the artfulness a teacher or instructional designer can muster will still be needed to make practice interesting and self-motivating. Drill and practice may succeed only to the extent that it fills obvious needs, is presented in palatable formats, and can be convincingly related to the broader subject matter of mathematics.

SUMMARY

We have presented an overview of the psychological research surrounding the use of drill and practice to build arithmetic skills. The associationist theory of E. L. Thorndike, applied to the classroom, has been used to justify drill as a means of forming and strengthening the stimulus-response bonds that are viewed as constituting the subject matter of arithmetic. Although increases in computational speed and accuracy appear to accompany most drill, various arguments have been advanced against using it as a principal method of instruction. The main objection is that drill cannot develop quantitative thinking because it treats mathematics as a collection of isolated bonds rather than an integrated set of patterns and principles. In contrast to drill methods, "meaningful" instruction in the number concepts underlying computations has the virtue of acknowledging differences in the level of understanding possessed by children and adults. It also appears to enhance the generalizability of arithmetic skills.

Various attempts have been made over the years to specify the sequence and organization of drill and practice in computation. Some have sought to ensure proper ordering and amounts of drill based on determinations of relative difficulty of number facts. Others have isolated structural variables whose summed strength predicts performance on word problems and thus serves as a means of ordering those problems according to difficulty. Studies comparing isolated and mixed forms of drill suggest that concentrated practice is best used in close proximity to the introduction of new facts or procedures, and small amounts of practice at frequent intervals are best for maintaining facts and procedures once they are well learned. Computer-assisted drill programs represent an attempt to optimize the effectiveness of practice by adjusting the timing and difficulty level to individual students. Supplementary drill in this individualized mode was shown to improve children's performance on standardized achievement tests but not more than supplementary paper-and-pencil drill.

Recent information-processing theories of human memory point to a possible theoretical justification for drill and practice—the development of automaticity. The capacity to respond automatically to certain components of complex computations, such as number facts and simple algorithms, may reduce the processing load of the human memory system and thus contribute to its efficient functioning. More precise indications of the value and effects of drill and practice are necessary in order to define its proper role in mathematics instruction.

REFERENCES

Brownell, W. A. *The development of children's number ideas in the primary grades.* Chicago: The University of Chicago, 1928.

Brownell, W. A. Psychological considerations in the learning and the teaching of arithmetic. *The teaching of arithmetic, the tenth yearbook of the National Council of Teachers of Mathematics.* New York: Teachers College, Columbia University, 1935.

Brownell, W. A., & Chazal, C. B. Premature drill. In C. W. Hunnicutt & W. J. Iverson (Eds.), *Research in the three R's.* New York: Harper, 1958. (Adapted and abridged from The effects of premature drill in third-grade arithmetic. *Journal of Educational Research,* 1935, *29,* 17–28.)

Brownell, W. A., & Stretch, L. B. *The effect of unfamiliar settings on problem solving.* Durham, N.C.: Duke University, 1931.

Buswell, G. T. *Summary of arithmetic investigations.* Chicago: University of Chicago Press, 1927.

Buswell, G. T. A critical survey of previous research in arithmetic. In G. M. Whipple (Ed.), *The twenty-ninth yearbook of the National Society for the Study of Education: Report of the Society's committee on arithmetic.* Bloomington, Ill.: Public School Publishing Co., 1930.

Clapp, F. L. *The number combinations: Their relative difficulty and frequency of their appearance in textbooks.* Bureau of Educational Research Bulletin No. 1. Madison, Wisc., 1924.

Craik, F. I. M., & Lockhart, R. S. Levels of processing: A framework for memory research. *Journal of Verbal Learning and Verbal Behavior,* 1972, *11,* 671–685.

Goals for school mathematics: The report of the Cambridge Conference on School Mathematics. Boston: Educational Services, 1963.

Huey, E. D. *The psychology and pedagogy of reading.* New York: Macmillan, 1908.

Hydle, L. L., & Clapp, F. L. *Elements of difficulty in the interpretation of concrete problems in arithmetic.* Bureau of Educational Research Bulletin No. 9. Madison, Wis.: University of Wisconsin, 1927.

Jacobson, E. *The effect of different modes of practice on number facts and computational abilities.* Unpublished manuscript, University of Pittsburgh, Learning Research and Development Center, 1975.

Jacobson, E. *The learning of number facts through computer instruction* (LRDC Publication 1976/25). Pittsburgh: University of Pittsburgh, Learning Research and Development Center, 1976.

Jerman, M., & Rees, R. Predicting the relative difficulty of verbal arithmetic problems. *Educational Studies in Mathematics,* 1972, *4,* 306–323.

Judd, W. A., & Glaser, R. Response latency as a function of training method, information level, acquisition, and overlearning. *Journal of Educational Psychology Monograph,* 1969, *60*(4), Pt. 2.

Knight, F. B., & Behrens, M. S. *The learning of the 100 addition combinations and the 100 subtraction combinations.* New York: Longmans, Green and Co., 1928.

Kramer, G. A. The effect of certain factors in the verbal arithmetic problem upon children's success in the solution. *The Johns Hopkins University Studies in Education. No. 20.* Baltimore, Md.: The Johns Hopkins Press, 1933.

LaBerge, D., & Samuels, S. J. Toward a theory of automatic information processing in reading. *Cognitive Psychology,* 1974, *6,* 293–323.

Lankford, F. G. *Some computational strategies of seventh-grade pupils* (Final report, Project No. 2-C-013). HEW/OE National Center for Educational Research and Development and The Center for Advanced Studies, the University of Virginia, October 1972.

Loftus, E. F., & Suppes, P. Structural variables that determine problem-solving difficulty in computer-assisted instruction. *Journal of Educational Psychology,* 1972, *63*(6), 531–542.

McConnell, T. M. Discover or be told? In C. W. Hunnicutt & W. J. Iverson (Eds.), *Research in the three R's.* New York: Harper, 1958. (Adapted and abridged from Discovery vs. authoritative identification in the learning of children, *University of Iowa Studies in Education,* 1934, *9*(5), 11–62.)

References

Miller, G. A. The magical number seven, plus or minus two. *Psychological Review*, 1956, *63*, 81-97.

Norem, G. M., & Knight, F. B. The learning of the one hundred multiplication combinations. *The twenty-ninth yearbook of the National Society for the Study of Education.* Bloomington, Ill: Public Schools Publishing Co., 1930.

Perfetti, C. A., & Hogaboam, T. The relationship between single word decoding and reading comprehension skill. *Journal of Educational Psychology*, 1975, *67*(4), 461-469.

Perfetti, C. A., & Lesgold, A. M. Coding and comprehension in skilled reading and implications for reading instruction. In L. B. Resnick & P. A. Weaver (Eds.), *Theory and practice of early reading* (Vol. 1). Hillsdale, N.J.: Lawrence Erlbaum Associates, 1979.

Repp, A. C. Mixed versus isolated drill organization. *The twenty-ninth yearbook of the National Society for the Study of Education.* Bloomington, Ill.: Public Schools Publishing Co., 1930.

Repp, A. C. Types of drill in arithmetic. *The tenth yearbook of the National Council of Teachers of Mathematics.* New York: Teachers College, Columbia University, 1935.

Suppes, P., Jerman, M., & Brian, D. *Computer-assisted instruction: Stanford's 1965-66 arithmetic program.* New York: Academic Press, 1968.

Suppes, P., & Morningstar, M. *Computer-assisted instruction at Stanford, 1966-1968: Data, models, and evaluation of the arithmetic programs.* New York: Academic Press, 1972.

Swenson, E. J. How to teach for memory and application? In C. W. Hunnicutt & W. J. Iverson (Eds.), *Research in the three R's.* New York: Harper, 1958. (Adapted and abridged from Organization and generalization as factors in learning, transfer and retroactive inhibition, *Learning Theory in school situations. University of Minnesota Studies in Education,* 1949, No. 2, 9-39.)

Tait, K., Hartley, J. R., & Anderson, R. C. Feedback procedures in computer-assisted arithmetic instruction. *British Journal of Educational Psychology*, 1973, *43*(2), 161-171.

Thorndike, E. L. *Educational psychology, Vol. II. The Psychology of Learning.* New York: Teachers College, Columbia University, 1913.

Thorndike, E. L. *The psychology of arithmetic.* New York: The Macmillan Co., 1922.

Thorndike, E. L. *The Thorndike arithmetics: Book three.* Chicago: Rand McNally, 1924.

Wheeler, L. R. A comparative study of the difficulty of the 100 addition combinations. *The Journal of Genetic Psychology*, 1939, *54*, 295-312.

3 TRANSFER HIERARCHIES AND THE ORGANIZATION OF INSTRUCTION

In the realm of mathematics, as in other fields, the school curriculum has classically followed a path from simple to complex—from adding and subtracting single digits to computing complex multidigit arithmetic problems to solving algebraic equations. In Chapter 2 we showed how developers of arithmetic drill programs have tried to build instruction on the basis of a progression from easier to harder problems. We also pointed out, however, that the early theories underlying drill were unable to explain why some bonds should be practiced before others or, in general, why any particular order of teaching should be better than any other. The result was an intuitive and empirical approach to deciding which problems should come where in a sequence—testing hundreds of problems on hundreds of children and then using speed or accuracy rates to order the content of instruction. Studies on problem difficulty, such as those of Brownell and Suppes (see discussion in Chapter 2, this volume), identified specific characteristics of problems—for example the number of computational steps or the size of the numbers involved. But even these studies were still essentially empirical in their approach. There was no theory to explain the psychological difficulty of various problem characteristics; thus, there was no explanation for why learning the easy problems first ought to make it easier to learn the harder ones.

To be a more powerful guide to instruction, an associationist theory—indeed any theory—needs a way of accounting for *transfer*; that is, it should be able to explain why "simple" learning assists more complex learning. A traditional account of transfer within associationist psychology held that successfully learning one task would make it easier to learn a second task to the extent that the two tasks contained some of the same components (i.e., the same sets of associations). This is known as the *identical elements* theory of transfer and has been explored in experiments beginning early in this century. Over time, the concept of identical elements was broadened to include rules and principles as well as individual associations (see Ellis, 1965, for background on theories of transfer). Although much of the early work on transfer was directly motivated by instructional concerns (for example, the work by Thorndike and Woodworth, 1901, and by Judd, 1908), experiments on transfer tended to be done in the laboratory and typically involved only two or three tightly analyzed tasks. Applying transfer theory to anything so complex as a school curriculum, therefore, represented a large leap from the transfer theories described in textbooks on the psychology of learning. An effort to make this connection as rigorously as possible has been made by Robert Gagné (1962, 1970), in what has come to be known as *cumulative learning theory*. This is a special version of an identical elements theory in which simpler tasks function as actual components (elements) of more complex tasks. The fact that complex tasks are made up of identifiable, simpler elements makes for transfer from simple to complex. This chapter describes cumulative learning theory and presents examples of Gagné's approach to analyzing skills into ordered subskills, or *learning hierarchies*. It considers how learning hierarchies can be validated and how they can enter into decisions concerning content, sequencing, and individualization of instruction. A final section introduces a special procedure called *rational process analysis* as a means of making overt the implicit assumptions about cognitive processes that underlie learning hierarchies.

LEARNING HIERARCHIES FOR MATHEMATICAL TASKS

It is easiest to develop the idea of a transfer hierarchy by looking at some examples. Figure 3.1 shows one of Gagné's hierarchies for a mathematical skill, adding integers (Gagné, Mayor, Garstens, & Paradise, 1962). Its target tasks represent two aspects of the same skill. Task 2 involves performing the actual computation. For example:

$$^-17 + (^-13 + {}^+62) + {}^+38 + {}^-100 = ?$$

Task 1 requires an explanation of the steps involved in formulating a definition of adding integers. The latter represents the sort of skill often valued by educators who stress comprehension of mathematical concepts as well as performance of computational routines.

A hierarchy is generated by considering the target task and asking, "What would [this child] have to know how to do in order to perform this task, after being given only instructions? [Gagné et al., 1962, p. 1]." The answer to this question for the tasks in Fig. 3.1 would be the subskills labeled Ia and Ib in the

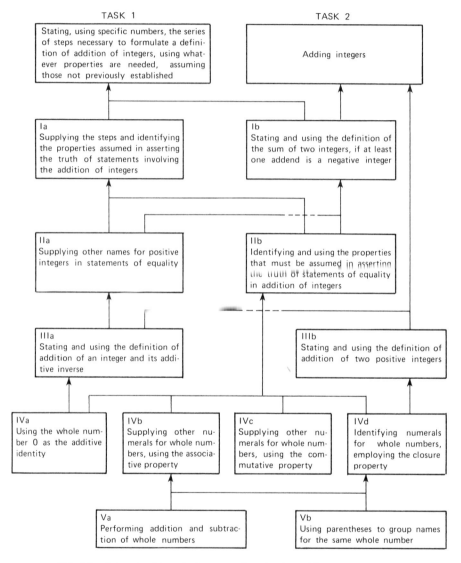

FIG. 3.1 A learning hierarchy for two tasks related to adding integers. (From Gagné, Mayor, Garstens, & Paradise, 1962. Copyright 1962 by the American Psychological Association. Reprinted by permission.)

second row. The upward-pointing arrows indicate that Ia is prerequisite to Task 1, and Ib is prerequisite to both Tasks 1 and 2. This means, according to the analysis, that Ia is a component of Task 1 so that it would be impossible to perform Task 1 without knowing how to perform Task Ia. In the same way, Ib is a component of both Tasks 1 and 2. Once a first set of prerequisite tasks is identified, the same question, "What would this child have to know . . . ," can be asked of the subskills. This is how the subskills labeled IIa and IIb were identified. They are prerequisites of Ia and Ib. The process can be reiterated until a complete hierarchy of successively simpler skills is generated. The hierarchy in Fig. 3.1 is meant to cover all cases of addition, simple or complex, involving both negative and positive integers, in equation or column form. For this reason, there are many subskills. The skills are also stated at a relatively high level of abstraction; they do not completely specify what the process of performance would be like.

A more limited capability, subtracting whole numbers of any size, is shown as a hierarchy in Fig. 3.2. This is drawn from the work of Gagné and Briggs (1974). According to this hierarchy, in order to subtract whole numbers, the child must be capable of "borrowing" under several conditions (Subskills VIII, IX, X, VII), subtracting without borrowing in successive columns (IV), subtracting one-digit numbers with borrowing (V), and so forth, down to being capable of giving answers to the basic subtraction facts (I). This last subskill could be further analyzed into simple discriminations, stimulus–response connections, and so on. However, the analysis does not usually go beyond the level of capabilities one might assume are already mastered by the student.

There are a few critical features and assumptions built into these hierarchy analyses that we should consider before going on to the question of how the analyses can be tested and validated. First, each of the identified skills and subskills is a performance capability, that is, something a person can *do*. In other words the skills are defined behaviorally. Gagné refers to the knowledge represented in hierarchies of this kind as sets of "intellectual skills," distinguishing them from factual knowledge such as memorized number facts or general understanding of mathematical structures and relations. Because hierarchies are defined behaviorally, one should assume only that their procedural components stand in a certain relationship to each other. The organization of general knowledge underlying these procedures can be very different, as we see in Chapter 8.

Second, the nature of a learning hierarchy is such that the subordinate tasks are *included in,* or *components of,* the highest-level task. This is only one way in which abilities acquired early in learning or development may influence later learning. Flavell (1972) suggests several kinds of transition relationships that may be needed to account fully for the development of competence in cognitive tasks. Early capabilities may become transformed into later ones; or ways of performing that were once useful but no longer are may simply drop from a child's repertoire. Of the several relationships suggested by Flavell, however,

42   3. ORGANIZATION OF INSTRUCTION

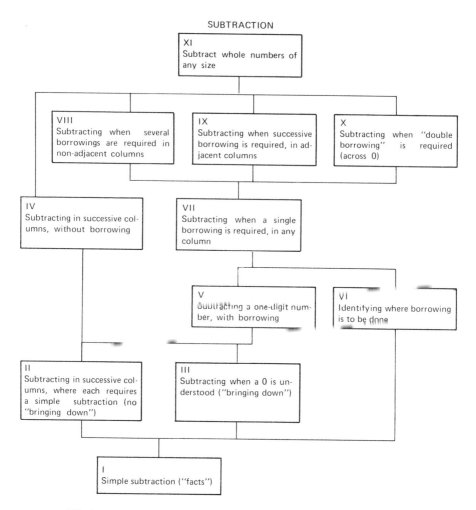

FIG. 3.2   A learning hierarchy for subtraction. (From Gagné & Briggs, 1974. Copyright 1974 by Holt, Rinehart and Winston. Reprinted by permission.)

inclusion of an early capability in a later one is considered to be of particular importance. Thus, although hierarchies may not account for all kinds of transfer relationships in development, they are useful as a way of conceptualizing learning in many domains, mathematics included.

Third, a task's position high in a learning hierarchy does not mean that it is harder to learn or that it will require more time and effort than each of the lower level tasks did. The higher-level tasks are indeed more complex than the ones below, because they are made up of all the component skills in the lower levels (plus some other skills that the analysis may not have identified). But the lower-order skills may in fact be the hardest to learn, in terms of time needed for mastery and in terms of the way thinking might have to be reorganized to accept the new skill. Likewise, the highest-order skill in the hierarchy may be easy to learn once all its components have been learned (Carroll, 1973); it may only involve adding a minor organizing feature to move from one level of skill to another.

Finally, each of the subskills identified in a particular hierarchy may also play a role in several other hierarchies. A learning hierarchy diagram usually includes the transfer relations among skills that enter into performance of a particular task. But this does not mean that those subordinate skills belong only to one hierarchy. Early prerequisite skills may be prerequisite to some other target task as well. Figure 3.1 provides a simple example; it is actually a diagram of two overlapping hierarchies, one for Task 1 and another for Task 2. The two tasks are highly interrelated, so they share many common components (e.g., Boxes Ib, IIa, & IIIa).

Just as particular components may be included in two very similar target skills, they may also show transfer to more distant domains of knowledge. For example, the skills acquired in learning how to read geographical maps may well be helpful to a person learning to interpret graphs. The new task may be learned much faster than the previous one, because so many components of the second have already been learned in connection with the first. These interrelations among tasks, based on transfer, are an integral part of cumulative learning theory, for they are the links that allow knowledge to "cumulate" and work together in the learning of new skills.

Validating Learning Hierarchies

It is not enough for a psychologist to take a skill, analyze it rationally into a learning hierarchy, and then hand it over to educators for use in designing a curriculum. A learning hierarchy is, at first, only a hypothesis about the way certain intellectual skills are related. As such, it is subject to testing, revision, and possibly even rejection. How certain can we be that a hierarchy of skills someone has developed really is a transfer hierarchy that will assist in learning and teaching? A number of methods have been applied to this problem. They fall into two general categories, scaling studies and training studies.

Scaling Studies. A common method for validating learning hierarchies is to test a group of students on each of the skills hypothesized as prerequisite to the target task. Each student scores either pass (+) or fail (−) on each component skill. Then students are arranged in an order such that the student passing most of the tests is ranked first and so on down the list to the student who passed the fewest. If the hierarchy being tested is indeed valid, that is, if the hypothesized transfer relations hold, then the scores should arrange themselves somewhat as those depicted in Fig. 3.3. This arrangement is called a *Guttman scale* (Guttman, 1944). If the scores form a near-perfect Guttman scale, it means that once we know the top-level skill that any individual student has attained, we know that the student is capable of performing all the skills subordinate to that one in the hierarchy. Student A, having passed Test 6, is also capable of Tests 1–5; Student E, having passed Test 2, can do Test 1 but not necessarily Tests 3–6. A pure scale such as the one shown in Fig. 3.3 is rarely obtained, of course. But there are statistical procedures available to help decide whether or not an actual set of scores can be considered to form a scale (see Lord & Novick, 1968, for a general discussion of scaling methods; see Resnick, 1973, and White, 1973, for methods specific to learning hierarchy validation).

If such a test does yield an identifiable scale, then we can be reasonably sure that we have identified a hierarchy of skills. If we gave Test 6 to a new student and if he or she passed it, then we could be quite confident that the student would also perform Tests 1–5 successfully. Similarly, a student who passed Test 4 could be expected to pass Tests 1–3; but if he or she failed Test 5, we would also expect failure on Test 6. If the scores based on a hypothesized hierarchy do not form a scale, the hierarchy is disconfirmed. In that case, we might try to reorder the skills in the hierarchy and try the scalability tests again to see if the reordered hierarchy was more valid. It might also be the case, however, that the test items did not accurately reflect the skills they were intended to measure. This could make a correct hierarchy appear invalid; conversely it might accidentally produce a scale that does not reflect a real relationship among the skills. For this reason it is necessary to accept scaling validations with considerable caution. In particular, one must be wary of accepting empirical scaling relations unless a relationship has been hypothesized in advance on the basis of a convincing rational

		Tests 1	2	3	4	5	6
Students	A	+	+	+	+	+	+
	B	+	+	+	+	+	–
	C	+	+	+	+	–	–
	D	+	+	+	–	–	–
	E	+	+	–	–	–	–
	F	+	–	–	–	–	–

FIG. 3.3 A perfect Guttman scale.

analysis. A rational analysis and an empirical scale that agree suggest that a hierarchy is probably valid. Either one alone does not constitute a validation.

An example of a scaling validation is a study by Wang, Resnick, and Boozer (1971). The study examined three basic classes of early mathematics behaviors: (1) counting objects; (2) comparing set size; and (3) using numerals. Using preschool children who had no previous instruction in number skills, the experimenters hoped to answer several questions: Do certain counting and numbering behaviors develop in a fixed order? At what point do children start thinking of numbers in terms of written numerals? How might the development of counting skills and of the concept of one-to-one correspondence be related to each other? A learning hierarchy was hypothesized to account for the relationships among these skills. Figure 3.4 shows the component abilities for counting and one-to-one correspondence and their hierarchical arrangement. Each box in the figure represents a specific task. The entry above the line describes the situation presented to the child; the entry below the line describes the desired response. For example, Box C is read, "*Given* a fixed, ordered set of objects, *the child can count the objects.*" The numeration skills, not depicted here, consisted of simple matching tasks (e.g., selecting a numeral to match a stated number) and the ability to read and write numerals. As the figure shows, counting and one-to-one correspondence were seen as developing independently, with neither skill being prerequisite to the other. Further, the hierarchy specified earlier development of these skills for smaller numbers (1–5) than for larger (6–10). The scores of children who had been given a battery of tests on these skills were tested for scalability. Most of the hypothesized relations in the study were confirmed. A strong finding was that counting tasks appeared to be prerequisite to learning and using written numerals. The implication for instruction was that whereas children could be taught to identify and read numerals as a rote–memory task, they would probably find numerals much easier to learn after having had a great deal of experience in counting objects. Since results of this kind are sensitive to small variations in test items, testing conditions, and the like, it is important to note that this finding was replicated in a follow-up study that used a different sample of children (Wang, 1973), thus adding a measure of assurance to its interpretation.

Independence of counting and one-to-one correspondence was also confirmed in both the original and the follow-up study, suggesting that these might be taught in either order. This is contrary to the beliefs of many mathematics educators who view one-to-one correspondence as prerequisite to counting because one-to-one correspondence is "built into" any number system. In many newer arithmetic programs, therefore, considerable time is given to correspondence between sets (embodying a cardinal definition of quantity), with counting and ordinal quantification being reserved for later. However, the findings of the Wang hierarchy studies are supported by quite a different line of research. Gelman & Gallistel (1978) have recently shown that preschool children

46 3. ORGANIZATION OF INSTRUCTION

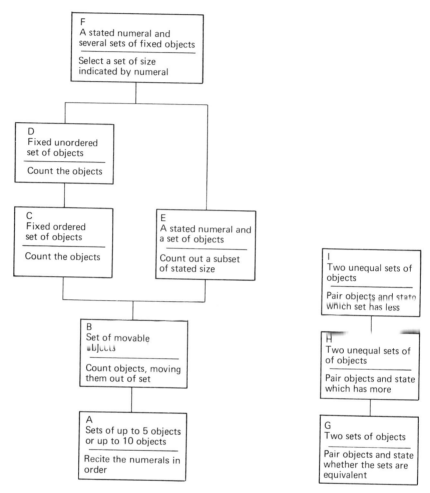

FIG. 3.4 Learning hierarchies for counting and one-to-one correspondence used in developing an early mathematics curriculum. (Adapted from Resnick et al., 1973.)

are able to count even before they demonstrate a grasp of one-to-one correspondence. Both the Wang and Gelman experiments are good examples of the way empirical psychological study serves as a check on formal theoretical analysis of a subject matter domain.

It is important to note the limitations of scaling studies for validating hierarchies that are developed with instruction in mind. The studies just described did not successfully validate the finer points of the hypothesized relationships among the skills. They were much better at confirming the general dependence of learning numerals on counting and the independence of counting from one-to-one correspondence. Several different kinds of counting tasks are described in Boxes A through F. These tasks did not scale in the way the hierarchy would predict—and in fact they scaled differently in different replications of the study (except that Task F was always at the top of the scale). This failure to validate predictions based on small differences between tasks is quite typical of hierarchy scaling studies. In general, this kind of research is capable of identifying the orders in which relatively large chunks of a curriculum can be effectively taught but not in suggesting detailed sequences within these chunks. (See Resnick, 1973, for a fuller discussion of this point.)

Another limitation of scaling studies derives from the fact that they assess only the *previous* learning of the experimental subjects. They therefore demonstrate only that under existing educational conditions one task is normally learned before the other. They cannot demonstrate that there is direct transfer from one skill to the next or that the order in which the skills are normally acquired is the most efficient, because there is no certainty as to how and in what order each skill was actually acquired. To draw conclusions of this kind, studies are needed that control the training that students receive. By directly training students in the components of a new skill they are to learn, an experimenter is better able to isolate the transfer effects of the hypothesized hierarchical ordering. This approach to validation is described next.

Training Studies. In a training validation study the hierarchy is used as the blueprint for an actual instructional sequence. Gagné has reported a number of such studies (Gagné, 1962; Gagné et al., 1962; Gagné & Paradise, 1961). We focus on one of them (Gagné et al., 1962) to show in some detail one way training studies have been used to validate hierarchies.

Gagné and his colleagues designed a programmed course of instruction based on a chapter out of a seventh-grade mathematics textbook. The subject matter of the chapter—the addition of integers, with a heavy emphasis on stating the principles and steps involved—was analyzed into the set of interrelated skills already shown in the learning hierarchy in Fig. 3.1. Instructional materials were prepared that would teach each of the subordinate skills outlined in the hierarchy, and a group of seventh graders went through the resulting learning program. The target skills themselves were not taught. Following training, the students were given a test of performance on the target tasks and a test on each of the subordinate skills that had been taught in the program.

The analysis of results proceeded as follows: The experimenters examined the relations between all pairs of skills that stood in an adjacent higher–lower relation in the hierarchy (i.e., pairs directly connected by arrows). The scores ("pass" or "fail") for each person tested could fall into four possible relationships, as shown in Fig. 3.5. The number of cases of each type of relationship was then counted for each skill, which gave an indication of how accurately the hypothesized hierarchy predicted transfer from the lower- to the higher-level skill. The experimenters found a large number of pass–pass (+ +) relations, a moderate number of fail–fail (− −) relations, very few pass higher–fail lower (+ −) relations, and a moderate number of fail higher–pass lower (− +) relations. These cases were from 97–100% in agreement with the theoretical predictions of their hypothesized transfer hierarchy (i.e., they showed the first, second, and fourth relationships of Fig. 3.5).

The data from this study were also analyzed another way, asking whether intermediate-level skills could be shown to facilitate transfer from skills below to skills above them on the hierarchy. Scores were compared on skills that were two levels apart in the hierarchy. Then the effect of either learning or not learning the intervening skill on the score for the higher-level skill was determined. When the intervening skills had been learned, the success rate on the higher-level task ranged from 44 to 89%; when the intervening skill had not been learned, the success rate on the higher skill ranged from only 0 to 33%. This finding too supported the hypothesis that learning subordinate skills helped to mediate the transfer from the simplest skills to the higher-level skills.

Although this study lent support to the hierarchy under test, it was far from a tight validation of the detailed sequences of learning that the hierarchy hypothesized. In the first place, measures that are based on a small number of test items (in this case as few as two per component skill) can be quite unstable. If the researchers had administered a large number of items for each skill in the hierarchy, at least a few students would probably have passed some and failed others; in other words, they would not have shown perfect pass or perfect fail patterns. The fact that each student was classified simply as passing or failing inevitably reduces the reliability of conclusions that can be drawn. Despite this methodological difficulty, the high level of agreement with hierarchy predictions is supportive of the hierarchy as a whole.

Higher	Lower	Which means:
+	+	Transfer occurred from lower to higher skill, as predicted by the hierarchy.
−	−	Transfer did not occur, also as predicted by the hierarchy.
+	−	Contrary to prediction that learning of higher skill is not possible without learning lower skill first.
−	+	Not contrary to predictions of hierarchy, but may show the learning program was weak.

FIG. 3.5 Possible relation between scores of tests of adjacent pairs of skills in a learning hierarchy. The symbols + and − denote *pass* and *fail*, respectively. (Adapted from Gagné et al., 1962.)

However, we could be more confident in the validity of the Gagné et al. (1962) hierarchy if transfer had been measured at a more detailed level—between component tasks themselves. As it was, the experiment *directly* tested transfer only for the target tasks. Instruction was given on all the skills leading up to these tasks; thus, we assume that those students who passed the test for Tasks 1 and 2 were helped by having learned the subordinate skills. The data bear this out: Of those who passed Task 2, adding integers, 74% had also passed Ia. But of those who failed the target task, only 4% had passed Ia. In other words, knowledge of subordinate skills, taken together, produced transfer to the target skills. However, we really know little about transfer among the subordinate skills themselves. This is because the subordinate skills were taught to all the students, and they were all taught in the same order (working *up* the hierarchy from simple to complex). There is really no way to tell how much of the learning was due to transfer from one skill to the next because there is nothing to compare it with. To estimate transfer with more accuracy we would want to know how much easier it was to learn a skill *after* learning its immediate subordinate than *before* learning the subordinate.

Another type of training study more directly tests the validity of a learning hierarchy by training a set of skills in different orders: Some individuals are taught in the order hypothesized by the hierarchy; others, in the reverse order. Using this method in teaching children how to compare sets, Uprichard (1973) uncovered a slightly different relation among set comparison skills than that predicted by a mathematical analysis. By teaching groups of preschoolers three concepts—equivalence (E), greater than (G), and less than (L)—in all possible orders, Uprichard confirmed that, at least under his instructional conditions, E was prerequisite to G and L, as expected. However, contrary to expectation, G and L did not seem to be simple complements of each other: L proved harder to master than G, as indicated by the longer time required to learn L when it preceded G in the instructional sequence. This suggested a teaching sequence in which the concept "greater than" precedes "less than," although a simple mathematical analysis might have led to teaching them simultaneously. Support for the generality of Uprichard's finding comes from developmental studies that show that the concept "more" is learned earlier than the concept "less" (Donaldson & Balfour, 1968; Palermo, 1974).

Using Hierarchies for Teaching

Clearly, it is appropriate to interpret the details of learning hierarchies with caution. Nevertheless it is also clear that hierarchies have enough psychological reality to justify using them to guide instructional planning. How, then, can learning hierarchies help us teach children the skills of mathematics? Once we know we have a learning hierarchy that predicts transfer with reasonable certainty, how can we use it?

Hierarchies as Individual Maps for Instruction. One application that immediately suggests itself is to use hierarchies as a map for a sequence of instruction. We can think of each defined skill as a task that one must be able to perform successfully. If several skills are related in a hierarchical way, then we can design instruction in each task and present the instruction in sequence. This should ensure that students progress smoothly through a sequence leading to successful performance of the topmost task in the hierarchy. For teaching, in other words, cumulative learning theory implies that any given task or curriculum objective can be broken down into simpler components. These components can be organized into a hierarchy, and one can expect transfer of learning from lower levels of the hierarchy to higher ones.

Resnick, Wang, and Kaplan (1973) designed an introductory mathematics curriculum according to these principles. Figure 3.6 depicts the various units covered by the curriculum and their hierarchical relationships. Each unit comprises a specified set of four to eight objectives, stated behaviorally and organized into hierarchies. The objectives for Units 1 and 2, for example, are laid out in Fig. 3.4, which we referred to earlier. (These objectives may also be used in interpreting the graphs of individual student progress to be described later.)

The curriculum was used as an individualized instructional program. Tests were prepared for each unit and for each task within a unit (Wang & Resnick, 1978). These tests were "criterion-referenced" (Glaser, 1963), which means that they sampled specific tasks in the curriculum hierarchy; they were not designed to estimate learning styles or general levels of ability or to compare individuals or groups with one another. The tests were used at the beginning of the school year to place individual children in the curriculum by determining which skills they had mastered and which they still needed to learn; later the tests were used to monitor individual progress. The hierarchical structure of the curriculum aided in the placement process, because it was not necessary to test each child on each skill. Instead, it could be assumed that a child who passed a unit or skill would not need instruction in any skills lower in the hierarchy. Conversely, a child who failed a skill would need instruction on that skill and also on skills at higher positions in the hierarchy. Individualized instruction began with the lowest skills not already passed in pretesting. Each instructional unit was followed by a posttest to determine whether the child had mastered the skill. In some uses of the curriculum, the child also took a pretest on the next skills in the hierarchy to see whether he or she still needed instruction on these skills or had acquired them through transfer.

Figures 3.7 and 3.8 display the progress of two kindergarten children, Marie and Roy, over the course of a school year. Objectives mastered during each month are marked with an *x*, yielding a relatively detailed picture of the pattern of mastery shown by each individual. The inset graphs are summaries that show the number of new skills mastered each month; these permit easier comparison of learning rates among children.

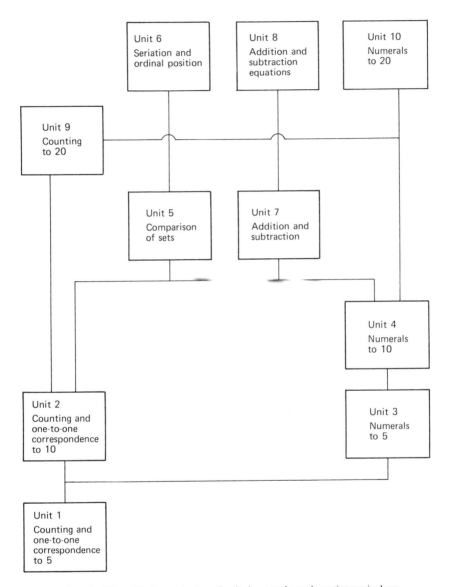

FIG. 3.6 Hierarchical organization of units in an early mathematics curriculum. Within each unit, specific objectives are also organized in a hierarchy. Fig. 3.4 details the objectives for Units 1 and 2. (Adapted from Resnick et al., 1973.)

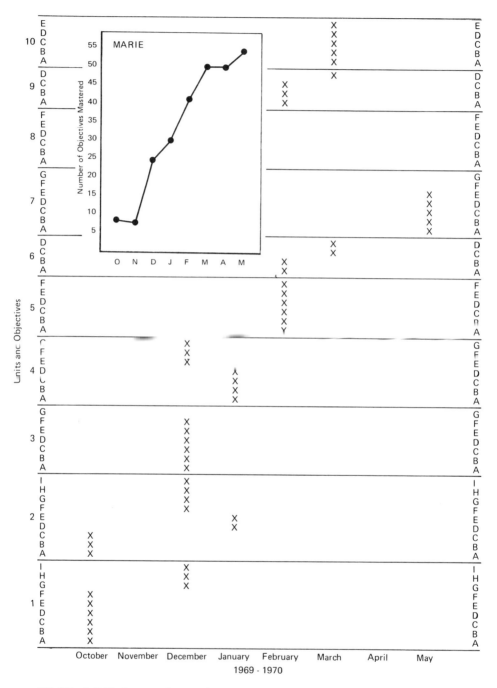

FIG. 3.7 Individual pattern of mastery in an individualized mathematics curriculum (Marie). (From Resnick et al., 1973. Copyright 1973 by Society for the Experimental Analysis of Behavior, Inc. Reprinted by permission.)

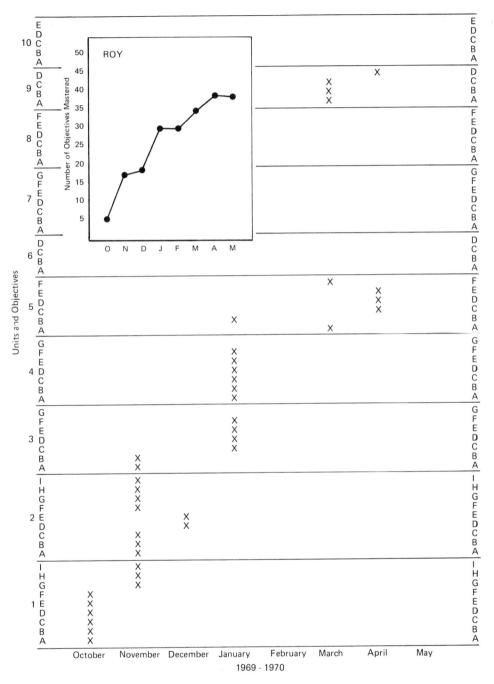

FIG. 3.8 Individual pattern of mastery (Roy). (From Resnick et al., 1973. Copyright 1973 by Society for the Experimental Analysis of Behavior, Inc. Reprinted by permission.)

The graphs illustrate the ways in which hierarchies can both reveal individual differences among students and help teachers adapt to them. Any group of children—even a group at the same general level of development—will differ in capabilities with respect to a specific mathematical task at any specific moment in time. The two children shown here, for example, differ somewhat in the number of skills mastered by October, and they differ markedly by December. The difference persists through the end of the school year. The sharp rise in skills Marie has mastered by December may reflect a real learning "spurt." But it may equally well reflect a classroom constraint: The teacher may not have found time to test Marie during November.

Despite such imperfections in practice, the use of hierarchies as a basis for determining entering capabilities means that one need not start with the simplest tasks of the curriculum in teaching every child. The diagnostic testing made possible by hierarchies permits teachers to locate individuals in terms of their skill levels and begin instruction at that point. In addition, as the graphs make clear, testing reveals the different rates at which children master skills, so that teachers can plan accordingly. To the extent that an identified hierarchy is valid, this method of teaching helps to ensure that every child is given the best chance of learning successfully. Instead of teaching in one way to all children, hoping that not too many will "drop out" of the learning sequence along the way, the teacher can help each child acquire the specific components needed to master the target skill.

Differences in Sequence of Instruction. A well-tested hierarchy establishes clearly certain transfer and prerequisite relations. But it does not completely specify the order in which a particular child should learn the tasks. When, for example, several skills are independent of each other (as are Units 2 and 3 in Fig. 3.6 or Skills C and E of Units 1 and 2 in Fig. 3.4), they can be learned in either order as long as they are both learned before the target task is attempted. *Which* order can be determined according to the individual child. Thus, even within a well-established hierarchy there are some branches and choice points, and teachers discover others in the course of classroom work. These are important in allowing alternative routes through a curriculum. Marie's and Roy's graphs (Figs. 3.7 and 3.8) illustrate the use of these options, in that the children sometimes work on several units at once and master Units 9 or 10 ahead of 5 or 7.

The graphs also demonstrate the extent to which even carefully researched hierarchies do not always describe necessary or optimal sequences of instruction. Both children mastered Skill F of Unit 2 before D and E, contrary to the hierarchy's specifications, and Roy mastered B before A in Unit 5. In the case of the curriculum portrayed here, repeated evidence of this kind was used to modify prerequisite specifications in its subsequent versions. But the finding is a general one; hierarchies have rarely, if ever, succeeded in specifying precisely the learning routes of individual children. For instructional purposes, then, a hierarchy provides a structured sequence for the teacher and student but not a completely determined one. There is room, indeed need, for judgment and adaptation to the interests of the child and teacher, even in the most fully validated hierarchies research has given us thus far.

Do Hierarchies Constrain the Able Learner? A more general question that is frequently raised about learning hierarchies, and about the cumulative learning theory on which they are based, is whether they imply that learning must always proceed carefully through small steps. We know that individuals do vary widely in how quickly they learn and that this often depends on whether or not they need explicit instruction on each skill in a hierarchy. A thoroughly individualized and adaptive approach to teaching mathematics has to be concerned, then, not only with a child's level of learning but also with the extent to which that child needs direct instruction in every step along the way.

It is evident that there are times when certain individuals are able to "skip" prerequisites, when they seem to learn complex behaviors without explicitly practicing the skills regarded as prerequisites for those behaviors. These occasions may arise, for example, when a child's motivation to learn is high, when he or she is lacking only a few prerequisites, and when the task is presented in a way that suggests the applicability of skills and concepts acquired elsewhere. There may be other occasions when all the component skills have been learned and need only to be combined and organized into a more complex new performance. In such cases, direct practice on the final task alone may produce dramatic learning effects, and it will appear that the child has skipped prerequisites in the hierarchy. This effect is not surprising, considering that there are so many factors besides direct instruction affecting the course of each child's learning, including skills learned in other domains.

There have, in fact, been a few studies in which psychologists taught only the highest-level task in a hierarchy and where children who managed to learn the target task *also* acquired the lower-level prerequisites as a by-product of this learning. Dienes (1963) was able to show incidental learning of simpler concepts in the context of learning complex games that involve the concepts of mathematical groups [closed sets of binary operations that have an identity element and an inverse for each operation (e.g., the set of possible rotations of a regular tetrahedron)]. Students who learned a mathematically more complex game first learned a simpler version of the game more quickly than those who learned the simpler game first, that is, the combined times for learning both games was less for students learning in the complex-to-simple order than it was for those learning in the simple-to-complex order. Dienes (Dienes & Golding, 1971) has suggested that certain mathematical concepts may need to be introduced in some measure of complexity rather than in the small sequential steps suggested by analysis into simpler components. The rationale for this so-called "deep end" approach is that the mathematical framework underlying certain concepts must be understood to make learning meaningful and to facilitate the learning of simpler components by providing an advance organizer. We examine this idea that certain organizing principles of mathematics may help children learn the details of particular computational procedures later in this book (see Chapters 5, 6, and 8).

Other studies, however, have suggested that the power of "complex-first" learning may be limited to certain kinds of students. A pair of training studies concerned with basic classification skills (Caruso & Resnick, 1972; Resnick, Siegel, & Kresh, 1971) showed that most students learned best when the skills were taught in the hypothesized hierarchical order but that a few were able to learn higher-level components without first learning the lower-level ones. However, all students who successfully learned the complex skill first, upon testing, showed that they had *also* mastered each of the prerequisites. This confirms the psychological reality of the skill hierarchy but suggests that some individuals are able to learn prerequisite components in the context of a more complex skill, without explicit instruction on each subskill. Such a finding perhaps signals a consistent pattern of individual differences that could serve as a basis for individualizing instruction. Under an individualized program, some children might proceed through rather tightly ordered sequences, whereas others might be encouraged to skip prerequisites and depend more on their own powers of inference. This potentially important dimension of individual difference in learning has not been systematically explored by psychologists. An extensive review (Cronbach & Snow, 1977) of research literature on the interaction between instructional programs and individual differences has revealed a recurrent finding: Highly "structured" teaching benefits students of low IQ or low general scholastic aptitude but has little effect on students of high aptitude. This research on aptitude-treatment interactions (ATIs) has not usually been conducted in the context of hierarchies, but the so-called "structured" teaching that benefits low-aptitude students usually means explicit instruction in a carefully sequenced set of steps. The general characteristics of most hierarchy-based instruction are thus present.

Science and Art in Applying Hierarchies. Data of this kind make it clear that in choosing a teaching strategy it is important to know a good deal about individual children's learning characteristics. Some children seem to learn in small steps, whereas others make larger leaps, acquiring the prerequisite skills almost incidentally. Testing based on hierarchies of mathematical tasks cannot in itself reveal which children are which. But as teachers observe children's behavior and begin to detect different patterns of learning, hierarchies can help them adapt to these individual characteristics. For some children it may be wise to teach each subordinate skill identified in the hierarchy. For other children, it may be better to teach in a way that deliberately encourages leaps of understanding that simply skip over many of the steps in the hierarchy.

Unfortunately there is little scientific basis at this time for making the individual decisions. The difficulties in validating learning hierarchies, along with the paucity of individual difference research in the context of hierarchically ordered curricula, suggest there is as yet no strong empirical basis in psychology for guiding teachers' observations. In particular, we do not know whether the ability to skip over component tasks is indeed a consistent trait of certain learners or whether it is tied to specific content or to particular learning situations. Gagné's summary of the state of learning hierarchy research in 1968 still holds today:

> A learning hierarchy... cannot represent a unique or most efficient route for any given learner. Instead, what it represents is the most probable expectation of greatest positive transfer for an entire sample of learners concerning whom we know nothing more than what specifically relevant skills they start with [p. 3].

Learning hierarchies, then, can be useful tools to help teachers and instructional designers make explicit their understanding of the organization of skill learning and the way individual children differ in the extent of their learning. But they must be used with caution. Indeed, most hierarchies that are proposed for guiding instruction have been subjected to no empirical validation at all. And even where validations have been systematically carried out, difficulties associated with test reliability, statistical decision procedures, and inadequate sampling of the curriculum range make it risky to place a great deal of faith in the details of the hierarchy. Used with flexibility and discretion, however, carefully developed hierarchies can be useful in assuring that all children, including the less able, master the essentials of school mathematics, especially computational skills.

INTRODUCTION TO RATIONAL TASK ANALYSIS

In the preceding sections of this chapter we define learning hierarchies and give some examples. We also suggest how psychologists can determine when a hierarchy exists—through scaling and other kinds of validation studies—and when it does not. And we discuss how hierarchies, once established, can define in a practical way the domain that is to be taught and how they can help match teaching to individual differences in initial skill and learning characteristics. Up to now, however, we have said little about *why* hierarchies exist, except for the rather general notion that knowledge and skill cumulate and that one capability is called upon in learning another.

To understand more fully how and why hierarchies work, we need to think about what mental actions people actually carry out when they perform the tasks

58   3. ORGANIZATION OF INSTRUCTION

that appear in a hierarchy. What do they do when they count, for example? What are the operations involved in counting, and what demands do these operations make on general psychological capabilities such as memory or spatial ability? Similarly, what is actually involved in performing a complex multiplication problem, what steps are carried out, and what knowledge does each of these steps depend on? There are two ways to go about answering these questions: One is to analyze the tasks logically and intuitively, using what psychologists and educators already know or have theorized about children's mental processing capabilities. This is called *rational* task analysis. This mode of analysis, based on logic and intuition, contrasts with *empirical* task analysis. Empirical analysis also asks what children do when they are engaged in mathematical tasks but answers the question by studying in detail the actual performance of children on those tasks. Our concern in the remainder of this chapter is with rational analysis of task performance. Chapter 4 explores in depth various methods of empirical task analysis.

Underlying every hierarchy, there is an implicit theory of the *processes* people engage in when they perform the tasks that appear in the various levels of the hierarchy. It is possible and useful to make these implicit theories of how tasks are performed more explicit. Explicit process analysis of computational tasks deepens our understanding of how children learn a set of capabilities and how instruction can better respond to their learning needs. To show the power of process analysis, we use as an example a very simple task of early arithmetic, counting. Apart from simply knowing how to reel off the numerals in proper order—"one, two, three . . ."—a number of specific counting tasks can be identified, each taught or assumed at the beginning of arithmetic instruction and each subtly different from the others. We refer the reader to Fig. 3.4, which shows the details of a portion of the introductory mathematics curriculum discussed earlier. This hierarchy was based on detailed rational process analyses of each of the skills involved (see Resnick et al., 1973). Focusing now on the counting tasks, we see that each makes somewhat different demands on perception, memory, and other basic processes. These differences are what led Resnick et al. to hypothesize the prerequisite relations displayed in the hierarchy.

One kind of counting task (labeled B in the hierarchy) occurs when we give a child, Catherine, a collection of objects and ask her to count the objects and tell us how many there are. We assume that Catherine can touch and move the objects at will, so we call this the *count movable objects* task. Figure 3.9 shows an analysis of this task in flow chart form. It describes the actual steps that a child performing the task might carry out. Following flow-charting conventions, the various steps are placed in a specific order by means of arrows, and decision points are shown as diamonds. At the outset the child has in front of her a set of objects. What she does first (A) is to move an object out of the set and say "one." Then she repeats the action, moving an object out of the set and saying the next numeral (B). Each time an object is moved and a numeral said, a "test" must be made of whether any further objects are left (C). If there are objects still in the set, the cycle continues (loop back to B). If not, the last numeral named is declared as the number of objects in the set (D), and the task is completed.

The ability to synchronize a move with saying a numeral is essential to the definition of counting. (See Gelman & Gallistel, 1978, for an analysis of counting that is similar to this presentation.) When we say "counting," we mean that each object is enumerated once and only once and that each is paired with a numeral name in an ordered and conventional sequence. The fact that Catherine can move the object out of the set is not important mathematically, but it is important psychologically. It gives her a way of visually and physically keeping track of which objects have been counted and which have not. It helps her meet the once-and-only-once criterion.

The situation is quite different, and quite a bit harder for some children, when the objects *cannot* be moved out of the set (Boxes C and D of the learning hierarchy in Fig. 3.4). Suppose, for example, that the objects to be counted are rows of pictures or dots on a page. In this case, Catherine must touch (not move) each object, say a numeral, then go on to the next object, and say the next numeral. There must be no repetition of either numerals or objects, nor should any be skipped, and the touching and counting must be synchronized. Thus the task is similar to *count movable objects,* but because in this new task the objects

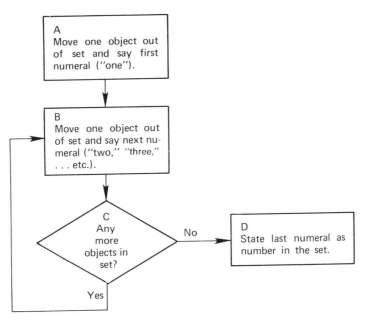

FIG. 3.9 Analysis of the count movable objects task. (Adapted from Resnick et al., 1973.)

60 3. ORGANIZATION OF INSTRUCTION

cannot be moved out of the set, Catherine has an additional burden. She has to keep track somehow of which objects have been touched without actually moving them. She touches them but leaves no trace of her touch behind. She must meet the once-and-only-once criterion entirely on the basis of some mental strategy for keeping track or "remembering" where she is in a spatial sequence. Thus there is substantially more cognitive demand on a child when the objects to be counted cannot be moved aside or otherwise marked off as they are enumerated.

The demand is perhaps not overwhelming if the objects to be counted are lined up in some neat fashion—a straight line perhaps or a number of rows. But what if the set were not organized spatially and were instead a collection of randomly scattered dots or other objects, such as a child might encounter in counting stars in the sky. Research on the processes by which children and adults manage such a task has shown that they visually "group" the objects and thereby establish a pattern for going through the set (Beckwith & Restle, 1966, discussed in Chapter 4). Each subset is either small enough or visually well-organized enough to allow the individual to keep track of which items have been counted. What the individual does, in other words, is create a series of organized sets out of what was presented as a single unorganized set. Then and only then can the person go ahead with the operations involved in counting an organized set. Even so, there is more chance for "getting lost" or forgetting an item than when the objects are actually arranged in some order. Figure 3.10 shows an analysis of the task *count fixed unordered arrays*. The memory requirements (keeping track of which objects and which subsets have already been counted) are not shown, but the figure helps to convey how complex the task is likely to be for a young child just learning to count.

We consider one more version of a counting task, one that, in fact, has been found to be considerably harder for young children than some of the others. This is a task in which a large collection of objects is given to the child with instructions to count out a certain subset of them (Box E in Fig. 3.4). The kindergarten teacher gives Paul a large box of buttons, for example, and asks him to give her 12 of them. Consider what Paul must do in this case. First of all, Paul must "store" in working memory the number he has been asked to count out. Once the target number is stored, counting can begin. This kind of counting is very much like counting movable objects (Fig. 3.9); it requires the same steps of moving an object out of the main set and saying the next numeral in the sequence. But there is a difference. Each time an object is counted, Paul must compare the stated numeral with the stored numeral and see whether it is time to stop (i.e., if enough have been counted out yet). If the most recently said numeral and the stored numeral do *not* match, then the operation continues. If the most recent and the stored numeral *do* match, then counting stops.

The need to keep the stored numeral in working memory and the recurring step of comparing the most recent numeral with the stored numeral appears to

Introduction to Rational Task Analysis   61

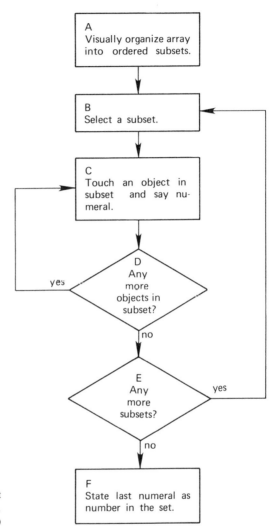

FIG. 3.10 Analysis of the count fixed unordered arrays task. (Adapted from Resnick et al., 1973.)

add a considerable load to the child who is performing the task. Unlike the test in Fig. 3.9, where there are more objects left to count, this is a comparison with something "in the head" and not with the way things look outside. The most frequent and obvious difficulty children have with this task is that they start counting out the originally presented set and keep going until there are no more left. In other words, they change the task into a *count movable objects* task. They simply omit the comparisons that would tell them when to stop.

Several difficulties might account for this behavior: It is possible that Paul "forgets" the target number in the face of all the competing mental activity that is going on. This is fairly likely when one considers the fact that he is constantly saying the numerals while counting; memory research (e.g., Underwood, 1957) has long established that more forgetting takes place when new items are similar to the ones to be remembered. Another possibility is that Paul retains the target number but fails to make the necessary comparison at each step. He forgets the task instructions or the "goal" of the task. Both kinds of errors can be accounted for by the limited capacity of working memory, as described in Chapter 2. The child who had trouble with this task may be concentrating on saying the numbers in the proper order rather than on keeping the target number in mind. As he or she grows a little older, reciting numbers in the correct order will become automatic, and the task will be easier because the more crucial processes will not be crowded out of working memory. Whatever the reason, there is evidence that counting out a subset of objects is more difficult for young children than simply counting a set (Wang et al., 1971; Wang, 1973).

The Uses of Rational Process Analysis for Instruction. These detailed examples of explicit rational analysis show differences in performance on tasks that on the surface appear very similar. Why go through the added work of such analysis? What does this technique offer the curriculum developer, and what does it offer the mathematics teacher in the classroom? First, as noted earlier, analyses of the processes involved in computational procedures are actually the basis of any curriculum design that attempts to order skills according to level of difficulty and particularly according to paths of positive transfer. One reason for carrying out detailed rational process analysis is quite simply to make the implicit analyses explicit. By detailing the mental steps we think are involved in each component skill, we improve the chances of developing learning hierarchies that do, in fact, facilitate learning.

For example, the process analyses just described suggest a specific sequencing for several kinds of counting activities. Counting movable objects makes fewer demands on memory than counting fixed objects. When fixed objects are organized in an ordered physical sequence, processing demands are fewer than when the objects are in a random array. With the random array, the individual has to impose and keep track of the order in which the counting operation is being performed. Rational analysis suggests that counting out a subset is the hardest task of all, because of the recurring need to test one's position in the counting chain against a stored numeral. Having analyzed the component tasks to such a level of detail, one has an extra measure of confidence in suggesting that instruction should proceed in the order indicated by the resulting learning hierarchy. What is more, by focusing attention on memory processes, the analyses point to specific suggestions a teacher can give a child who is learning to do the tasks. Where memory load is particularly taxing for a young child, as in counting out a subset, the child can be shown various place-keeping methods or memory aids, such as rehearsing the target number or writing it on paper.

Of course, the training sequence derived from any rational analysis requires validation through scaling studies, training studies, and/or detailed observations of task performance. And, as we pointed out earlier, fine-grained hierarchies such as the one in Fig. 3.5 are extremely difficult to validate. Children are sensitive to small and unintended differences in physical displays or verbal instructions, and they often invent ways of performing tasks that allow them to bypass some of the memory and other cognitive demands. (Resnick et al., 1973, describe some of these variant strategies; much more attention to children's invented strategies appears in Chapters 4 and 8 of this book.) Yet even if complete empirical validation of training sequences is not possible, rational process analyses can serve as valuable diagnostic tools for the teacher. Suppose that a child is having difficulty with counting, for example. In the absence of a process analysis that suggests what *should* happen in an efficient performance, there is little the teacher can do except provide more drill and practice on the same problem. With an analysis of the counting process in hand, even a hypothetical one, the teacher can try to figure out, by observing and questioning, whether the child is performing the task in the expected way at all and, if so, what particular aspects of it may be causing difficulty. Sometimes, having located the probable source of difficulty, a teacher can simply explain or demonstrate to a child what needs to be done. At other times, it may be helpful to use related tasks that include the same important processes but in simpler contexts. In any case, with a process analysis in mind, the teacher can become a keener observer, a more refined diagnostician, and on this basis a more skillful adapter to individual children's understandings and misunderstandings.

We have been touching already on the subject matter of Chapter 4—information-processing analyses of computation—in which we probe more deeply into the processing requirements of computational tasks. However, in Chapter 4 our focus is on empirical process analysis rather than on logical or intuitive analysis of tasks. The rational analyses we have presented here are process analyses that take into account as much as is already known about psychological limitations and capacities, but they do not claim to be tested, empirical descriptions of what people *actually* do. They are logically sufficient analyses; they suggest procedures or routines that are capable of solving computational problems. But these may not be the routines that people who are highly skilled actually use. Because of their rational character, therefore, the analyses presented thus far have clear weaknesses as psychological theories of mathematical behavior. As tools for instruction, however, they can be quite powerful. Almost anyone who understands the subject matter well and who has taken time to observe children performing the tasks involved can carry out rational task analysis. Two individuals' analyses may not agree in every detail, but the fact of having carried them out ensures a basis for explicit discussion of different hypotheses as to what goes on during computation or other mathematical performances. Just as important, a reasonably good rational task analysis can be performed by the teacher of mathematics and then used—in the ways described previously—to guide in diagnosing difficulties and adapting instruction to individual student characteristics. As empirical studies increase our knowledge of actual performance routines, these uses of rational task analysis can become increasingly tied to well-established knowledge about the psychology of mathematical learning and performance.

SUMMARY

This chapter has explored a method for organizing mathematics teaching based on a theory of how transfer occurs in learning. Cumulative learning theory views all subject-matter learning as a cumulation of increasingly complex elements, proceeding from simple stimulus–response connections through concepts and rules to higher-order problem solving and thinking. On the basis of cumulative learning theory, mathematical tasks can be analyzed into hierarchies of component skills that show positive transfer to higher-level skills in the hierarchy. The transfer relations may be verified by testing students on hypothesized subskills and judging whether or not the student performances can be said to form a Guttman scale. Additionally, training studies validate hierarchies by teaching specific component skills and verifying their transfer to higher-level skills in the hierarchy. Caution is needed in interpreting validation studies, however, because methodological variations and apparently small differences in statistical treatments can lead to different decisions.

Learning hierarchies suggest an order for instruction on component skills and also ways of individualizing instruction. Diagnostic testing based on a hierarchy permits teaching at each child's level of skill. Taking alternative routes through a hierarchy and skipping over certain prerequisites are some of the ways teaching can match specific children's learning characteristics. Some children need direct instruction in all the steps of the learning hierarchy and others seem to profit by skipping steps. The teacher can use a learning hierarchy as a basis for decision making that tailors instruction to individual differences. But artful intuitive application is required as many important questions about individual differences and transfer remain unanswered by psychological research to date.

The processes involved in tasks in a learning hierarchy can be analyzed explicitly as a further means of understanding children's learning and performance. Rational task analysis takes into account what is known and what is intuited or logically deduced about human information processing and task performance. It contrasts with analysis of actual performance on tasks, or empirical process analysis. Rational analysis is a useful tool both for making explicit the assumptions of mathematics curricula and for diagnosing learner difficulties.

REFERENCES

Beckwith, M., & Restle, F. Process of enumeration. *Psychological Review*, 1966, *73*(5), 437–444.
Carroll, J. B. Discussant comments. In L. B. Resnick (Ed.), Hierarchies in children's learning: A symposium. *Instructional Science*, 1973, *2*, 311–362.
Caruso, J. L., & Resnick, L. B. Task structure and transfer in children's learning of double classification skills. *Child Development*, 1972, *43*, 1297–1308.
Cronbach, L. J., & Snow, R. E. *Aptitudes and instructional methods: A handbook for research on interactions*. New York: Irvington, 1977.
Dienes, Z. P. *An experimental study of mathematics-learning*. New York: Hutchinson & Co., 1963.
Dienes, Z. P., & Golding, E. W. *Approach to modern mathematics*. New York: Herder and Herder, 1971.
Donaldson, M., & Balfour, G. Less is more: A study of language comprehension in children. *British Journal of Psychology*, 1968, *59*, 461–471.
Ellis, H. C. *The transfer of learning*. New York: Macmillan, 1965.
Flavell, J. H. An analysis of cognitive-developmental sequences. *Genetic Psychology Monographs*, 1972, *86*(2), 279–350.
Gagné, R. M. The acquisition of knowledge. *Psychological Review*, 1962, *69*(4), 355–365.
Gagné, R. M. Learning hierarchies. *Educational Psychologist*, 1968, *6*(1), 1–9.
Gagné, R. M. *The conditions of learning* (2nd ed.). New York: Holt, Rinehart & Winston, 1970.
Gagné, R. M., & Briggs, L. J. *Principles of instructional design*. New York: Holt, Rinehart & Winston, 1974.
Gagné, R. M., Mayor, J. R., Garstens, H. L., & Paradise, N. E. Factors in acquiring knowledge of a mathematical task. *Psychological Monographs: General and Applied*, 1962, *76*(7, Whole No. 526).
Gagné, R. M., & Paradise, N. E. Abilities and learning sets in knowledge acquisition. *Psychological Monographs: General and Applied*, 1961, *75*(14, Whole No. 518).
Gelman, R., & Gallistel, C. R. *The child's understanding of number*. Cambridge, Mass.: Harvard University Press, 1978.
Glaser, R. Instructional technology and the measurement of learning outcomes: Some questions. *American Psychologist*, 1963, *18*(8), 519–521.
Guttman, L. A basis for scaling quantitative data. *American Sociological Review*, 1944, *9*, 139.
Judd, C. H. The relation of special training and general intelligence. *Educational Review*, 1908, *36*, 42–48.
Lord, F., & Novick, M. *Statistical theories of mental test scores*. Reading, Mass.: Addison Wesley, 1968.
Palermo, D. S. Still more about the comprehension of "less." *Developmental Psychology*, 1974, *10*(6), 827–829.
Resnick, L. B. (Ed.). Hierarchies in children's learning: A symposium. *Instructional Science*, 1973, *2*, 311–362.
Resnick, L. B., Siegel, A. W., & Kresh, E. Transfer and sequence in learning double classification skills. *Journal of Experimental Child Psychology*, 1971, *11*(1), 139–149.
Resnick, L. B., Wang, M. C., & Kaplan, J. Task analysis in curriculum design: A hierarchically sequenced introductory mathematics curriculum. *Journal of Applied Behavior Analysis*, 1973, *6*, 679–710.
Thorndike, E. L., & Woodworth, R. S. The influence of improvement in one mental function upon the efficiency of other functions. *Psychological Review*, 1901, *8*, 247–261.
Underwood, B. J. Interference and forgetting. *Psychological Review*, 1957, *64*, 49–60.
Uprichard, A. E. The effect of sequence in the acquisition of three set relations; an experiment with preschoolers. In L. B. Resnick (Ed.), Hierarchies in children's learning: A symposium. *Instructional Science*, 1973, *2*, 311–362.

Wang, M. C. Psychometric studies in the validation of an early learning curriculum. *Child Development,* 1973, *44*(1), 54-60.

Wang, M. C., & Resnick, L. B. *The Primary Education Program* (12 vols.). Johnstown, Pa.: Mafex, 1978.

Wang, M. C., Resnick, L. B., & Boozer, R. F. The sequence of development of some early mathematics behaviors. *Child Development,* 1971, *42,* 1767-1778.

White, R. W. Learning hierarchies. *Review of Educational Research,* 1973, *43,* 361-375.

4 Analyses of Performance on Computational Tasks

In Chapters 2 and 3, we alluded to research and theoretical formulations that attempt to take us beyond a characterization of overt mathematical performance to some understanding of the thinking processes that occur as people work. We described rational analyses of computational tasks that try to lay out ideal sequences of mental routines for performing those tasks. But we did not consider whether those routines are actually used by people, nor did we outline the methods that psychologists might use for studying that kind of question. We turn now to *empirical* task analyses, analyses that attempt to determine exactly what goes on in a person's head from the time a mathematical task is presented to the time a response is given. To understand what empirical analysis entails, let us start with some intuitions about how the mind works.

A young student, Freddy, sits down at his desk to calculate the square root of 2798. What happens? We can see some of what he starts with: a paper and pencil, an assignment to find the square root, a number whose root is to be found. Freddy will generate an answer, either correct or incorrect, that we will also be able to see when it is recorded on his paper. This answer will be our evidence that some kind of processing occurred that transformed the given number into another number that seemed to satisfy the definition of the task. Of what might this processing consist?

We can guess that one of the first processing steps is for Freddy to recognize the configuration of lines on his paper as the number 2798, one of a set of numbers in his experiential repertoire. He will also perceive and attach meaning to the written or oral instructions, which will then call up out of his memory an algorithm he has previously learned for computing square roots. Freddy might then set himself a goal: Given a number, 2798, and an appropriate algorithm for computing the square root, apply the algorithm to the given number. The computation process itself will call up additional information from memory, such as specific number facts, formatting conventions, estimating techniques, and rounding procedures. So far, we have guessed that this task involves perception, memory, and problem-solving strategy. If all goes well, Freddy's processing efforts yield the correct answer. Comparing Freddy to a calculating machine, we might say he inputs certain information, processes that information, and produces an output. We can see both the input and the output. What we cannot see is the actual processing, that is, what Freddy does in the course of solving a mathematical problem, be it simple and algorithmic or complex and heuristic.

How does one obtain information about human information processing when the cognitive aspects of performance are not visible? Actually, there is much to be learned by examining the outward manifestations of thinking. We often assume we know how children perform arithmetic calculations, but our assumptions are based on our own methods or perhaps upon some notion of what we think they *should* do. What they *actually* do may turn out to be surprising. It is thus incumbent upon educational and psychological researchers to study human performance and on that basis to develop general models of performance to which individual performances can be compared. Psychologists have developed experimental methods directed specifically at interpreting observable behavior in terms of mental processing. Because of their concern with the systematic manipulations of data in the human mind, these psychologists have become known as information-processing psychologists. Often these researchers study skilled performance, that is, the behavior of a person who is considered to have mastered the skills required in a specific task. For those directly concerned with instruction, the models derived from this kind of study provide a standard of performance toward which children's learning can be guided.

The studies we describe in this chapter analyze the actual behavior of children and, in some cases, adults on computational tasks. We look first at behavioral data that allow us to make inferences about the processing demands of counting. This skill appears to be prerequisite to mathematical understanding, and it remains an important component of algorithmic computation for many children throughout the elementary school years. Then we present psychological models of the processing required in simple addition and subtraction tasks, based on time data. Detailed observations of children's performances on a set of basic arithmetic problems are reported, and the processing strategies characteristic of good and poor solutions are compared. We also describe ways in which computational errors can be analyzed in order to find systematic deviations from taught algorithms and to provide insights into processing steps. Finally, we show how psychologists have programmed a computer to solve algebra word problems and how, by comparing the computer's behavior with that of human subjects, they have helped to focus attention on the conceptual and representational aspects of problem solving.

SIMPLE MATHEMATICAL TASKS: THE USE OF REACTION-TIME DATA IN STUDYING PERFORMANCE

Since each step in a thought process takes a finite amount of time, it becomes of interest to study *how long it takes* to perform a task. By comparing the times required for two similar tasks, for example, one can infer differences in the number of steps and sometimes the types of steps involved in those tasks. Typically, the psychologist measures the length of time that elapses between the presentation of a problem and the subject's response, or reaction. The patterns of *reaction times* provide clues to how the mind interprets and responds to any given task.

The studies discussed in this section analyze data on reaction times as a means of evaluating hypotheses about how people perform relatively simple mathematical tasks. We deal in turn with the processing requirements of simple counting behavior and of addition and subtraction. The reader will note the minute level of detail in these analyses. The data were collected on special machines that record reaction times in milliseconds, and comparisons among individual performances are couched in terms of these very small fractions of seconds. But because the mind works at such a tremendous rate, even such tiny variations allow important conclusions about processing.

Determining Quantity: Counting and Subitizing

We begin by considering what are almost certainly the most elementary mathematical operations: counting and related ways of determining quantity. Recent research reported by Gelman and Gallistel (1978) tells us that children engage in quantification operations well before school age and that these operations are the building blocks of the more complex mathematical performances that are to follow. Although most children enter school knowing the basics of counting and quantifying, it appears that the skill develops with age and practice. It is thus worth considering in some detail the research on these very elementary mathematical operations.

By mathematical definition, counting is a process in which the objects in a set are noted one at a time, each object being noted once and only once. Further, as each object is noted, it is paired with a word (a number name) and these words are named in a fixed order ("one, two, three, ..."). This quantification process can be seen overtly sometimes, but often it proceeds silently. Does something like "internal counting" go on in such cases? If it does, then it should be true that the more objects there are in a set, the longer it will take to quantify (count) the set. A study by Beckwith and Restle (1966) confirmed this hypothesis. They found a systematic relationship between the number of objects and time, with each additional object to be counted adding a predictable amount of extra time (approximately ⅔ second for children and ⅓ second for adults) to the total time needed.

Counting objects one at a time works well where the number of objects is relatively small. Where large sets have to be quantified, however, one-by-one counting is not very efficient and there is a good chance of error. It is easy to lose one's place and either skip one of the objects or count it more than once. Beckwith and Restle (1966) hypothesized that in this situation skilled counters first group the objects into perceptually distinct subsets, then count each of the subsets, and finally cumulate the quantities. To investigate this hypothesis they varied the spatial organization of arrays of dots that they presented for subjects to count and examined the effect of different perceptual groupings on the time needed to count. They found that rectangular arrays (especially size 4 × 4) were counted particularly fast by both children and adults. The pattern of responses suggested that children used the rectangular format mainly to help them order and keep track of the objects to be counted; but adults appeared to make special use of the rectangular array by multiplying rows and columns, thus reducing the number of counts needed.

To investigate the effects of grouping further, Beckwith and Restle (1966) gave college students sets to count in which the positions of objects were totally scrambled, but in which color and shape cues sometimes functioned as aids to perceptual grouping. For example, perceptual grouping was strongly cued when four colors and four shapes were arranged so that in one section of the page there were all blue squares, in another all red circles, and so on. On the other hand, grouping was absent or weak when colors and shapes were scattered independently and randomly over the field. As might be expected, the better the color–shape cues for perceptual grouping, the faster the counting. Schaeffer, Eggleston, and Scott (1974) found a similar facilitation for children as young as 4 years.

Comparing Two Modes of Quantification. Counting is the dominant but not the only way we quantify. This may be verified through a simple demonstration. In Fig. 4.1 are pictured several arrays of dots. One need only glance at some of the arrays to determine how many dots they include. Counting is unnecessary for the arrays of 3 and 4 dots because the eye takes in their "threeness" or "fourness" immediately. This is called *subitizing,* a technical term that comes from the Latin root *subito,* meaning suddenly or immediately. Arrays of 3 or 4 dots are easily subitized, but as the number of dots increases to 6 or 10, this immediate recognition of quantity is no longer possible.

Klahr and his colleagues (see Klahr & Wallace, 1976) have studied the relation between subitizing and counting processes in children and adults. Their basic questions were at what point arrays become too large for people to subitize and whether there are changes with age in the ability to subitize. Klahr showed subjects patterns of dots, with from 1 to 10 dots in each array, and asked the subjects to report how many dots they saw each time. A special microphone

Simple Mathematical Tasks 71

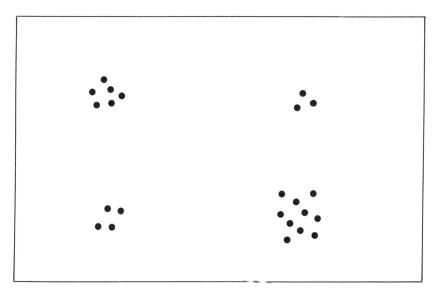

FIG. 4.1 Subitizing and nonsubitizing arrays. The arrays of 3 and 4 dots can be quantified without counting.

activated a voice-operated relay, which in turn operated a timer that precisely measured the reaction time on each trial.

Figure 4.2 shows three patterns of data one might expect to result from such a study. The ordinate (vertical axis) on each graph shows reaction time, measured in milliseconds—the higher the line, the longer the reaction time; the abcissa (horizontal axis) represents the number of dots in the array or, more generally, the number of objects to be quantified in a set. Figure 4.2a shows what the plotted line might look like if there were only a single way of quantifying arrays of any size, namely, straight counting. Reaction time increases in a single slope, continuous from very small to very large numbers of dots. Figure 4.2b indicates how the plotted reaction times might look if there were such a phenomenon as "pure" subitizing, that is, if it actually takes no more time to quantify relatively large arrays (e.g., 5 objects) than relatively small ones (e.g., 2 objects). Where subitizing ends, counting takes over and from there the graph shows a sloped line indicating a steady increase in time with each additional object counted. A third possibility, shown in Fig. 4.2c, is that within a certain range of quantities of dots there is a slight slope, but the slope changes sharply at the end of that range. This would suggest that, even when subitizing, extra time for scanning each additional object is needed but that this time is slight compared to what it would take to count each object.

Klahr's first experiments used adult subjects, and the data looked most like Fig. 4.2c; that is, there was a very shallow slope up to three or four dots, and then

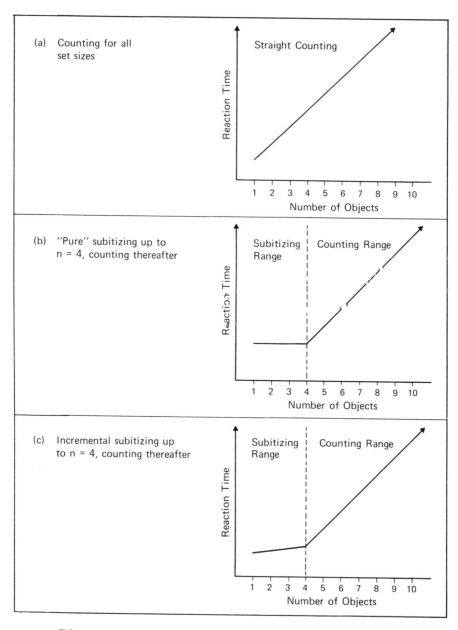

FIG. 4.2 Hypothetical data demonstrating subitizing and counting processes. Plotted lines show increases in total reaction time with each additional object to be quantified. In adults, counting averages 300 msec per object; subitizing averages 50 msec.

a steep one for larger arrays. The addition of each dot to an array up to three or four dots increased the reaction time by 50 msec or so. Above that number, each new dot required about 300 msec, which is the speed of silent counting in adults, as determined by Beckwith and Restle (1966). These data support the hypothesis that the subjects were able to subitize very small arrays but had to count larger arrays one by one.[1] Even under several variations in experimental conditions, the slopes for the subitizing range never went over about 70 msec per dot, and there was always a sharp increase in time per additional dot above 3 or 4 at which point, it was assumed, subitizing became impossible and counting took over.

Subsequent studies used the same method to examine subitizing and counting rates in 5-year-old children (Chi & Klahr, 1975). The basic pattern of Fig. 4.2c continued to hold, that is, for the smallest size sets there was a shallow slope, indicating subitizing, although for larger sets the slope was steeper, indicating counting. But the children took four times longer for each extra item in the subitizing range than did adults (about 200 msec as compared with the adults' 50); and they took longer for each additional item in the counting range (over 1000 msec, as compared with 300 for adults).

Klahr and Wallace (1976) have argued that subitizing is not only faster than counting, but it also develops earlier and is in fact prerequisite to the development of counting in children. These differences in quantification modes have implications for other aspects of children's mathematical development. The implications can best be considered with reference to a memory model, described in Chapter 2, in which the processing capacity of working memory is limited. In general, we can assume that subitizing takes up less of the limited space in working memory than does counting. It is more of a "chunk," less a set of separate operations. This being the case, it should be easier for people to use information about quantity if they have obtained that information through subitizing than through counting: There will simply be more room left in working memory for other necessary operations.

Consider, for example, the problem of conservation of quantity (discussed in greater detail in Chapter 7). To conserve means to realize that changing the physical arrangement of an array does not change its quantity. It means being able to hold in mind simultaneously the initial arrangement and its quantity and the transformed array and its quantity. To acquire conservation, one must have made this comparison enough times to be able to "trust" one's prediction that a change of physical arrangement will not affect quantity. It seems reasonable that one could do this more easily for arrays small enough to be subitized than for larger arrays that needed to be counted. In other words, children should demonstrate an understanding of conservation if quite small sets are used, but they may

---

[1]An alternative hypothesis that fits Klahr and Wallace's (1976) data is that the large arrays were quantified by subitizing subsets and then adding the subset quantities. It seems likely that adults use this latter procedure more than children.

74   4. ANALYSES OF PERFORMANCE ON COMPUTATIONAL TASKS

not conserve if larger sets are used. Although the data on conservation behavior are not yet substantial enough to prove the case, they certainly suggest that, in general, concepts involving quantity relations develop first in situations where the task of establishing quantity is itself simple—perhaps only a single chunk in short-term memory rather than a several-step procedure (such as counting).

The notion that subitizing precedes counting in the course of development or that it is more basic to mathematical understanding than counting has been challenged by Gelman and Gallistel (1978). They present evidence that 3-year-olds show spontaneous, skilled counting behavior with small sets of objects and that many 2-year-olds already understand and try to apply certain components of counting rules. But Gelman and Gallistel agree that quantification in young children is well developed only for small sets and that other forms of mathematical reasoning (conservation, for example) depend on secure quantification. Thus, we think they would agree with the instructional suggestion we draw from the work on early quantification: If one wants to teach a concept that depends on knowing the cardinality of sets (e.g., the concept "greater than" or "less than"), and if it is the *concept* that matters rather than the size of the set to which it applies, then one should perhaps let children work at first with very small sets. Similarly, since counting appears to become more efficient with age and, presumably, with practice, it may be well to delay conceptual work, or even advanced computational work, on large sets until counting skill is well enough developed to proceed at the rate of about three items per second instead of the one per second that seems to characterize 5-year-olds.

Solution Processes for Simple Equations

We turn next to studies of some of the basic tasks of school arithmetic. Our focus is on the processes by which children compute the answers to simple addition and subtraction problems involving single digits ranging from 0 to 9 (e.g., 3 + 6 = □, 9 − 6 = □). If an adult were asked "How much is 3 + 4?" he or she would probably know immediately without really having to figure out the answer. Most adults have stored in long-term memory a response, 7, that is linked with the stimulus 3 + 4. It is as if there were a huge directory in their heads, and some of the entries were number facts that merely had to be "looked up." But think, now, what happens when a person has an occasional lapse of memory, when a number fact slips out of grasp. The answer is usually reconstructed in some way. This reconstructing process is close to what young children do most of the time. Analysis of reaction time data allows us to study such processes as adding and subtracting even though they are ordinarily carried out internally.

Hypothesized Models of Addition. Let us look at three models of performance on addition equations that were hypothesized by Suppes and Groen (1967; Groen & Parkman, 1972). Because counting appears integral to the processing of

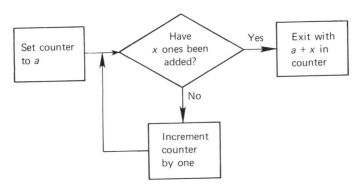

FIG. 4.3   Model of a mental counter that is set to a quantity, *a*, to which another quantity, *x*, is added by increments of one. (From Groen & Parkman, 1972. Copyright 1972 by the American Psychological Association. Reprinted by permission.)

number facts in young children, the models have in common the notion that there is some sort of "counter" in the head that can be set to some initial quantity and then "tick off" each increment as counting proceeds. Fig. 4.3 schematizes the process. First the counter is set to a quantity, *a*. Then the incrementing loop begins. Before each increment—which is always by one—a test is made to see if the specified number of increments (x) has already been made.[2] When the test yields a positive answer, processing exits from the incrementing loop and the counter can be "read." To compute the answer to the problem 3 + 4 = □, several versions of the Fig. 4.3 sequence are possible.

One model suggests the following processing sequence: The counter is set to 0 and then incremented three times; then, without resetting, it is incremented four more times. The final output, 7, can now be "read" on the counter. This is a reasonable way of adding and it is actually close to the procedure for adding that children are first taught in school.

A second model of addition is a bit more sophisticated. It still assumes a counter in the head, but this is set not at zero, as in the previous model, but at whatever the first number is in the equation. So, for the problem 3 + 4, the counter is set to begin counting from the number 3. It then increments four times, again yielding a readout of 7. This is a more sophisticated procedure because the person doing it recognizes that the number 3 always stands for the same quantity. The quality of "threeness" need not be verified each time by counting to 3.

---

[2]In a full specification of this model there would have to be two counters—one that is set to *a* and incremented from there, and a second that is always incremented from zero in order to keep track of when *x* increments have been made. However, since the second, "keeping track" counter would always make exactly the same number of increments as the main counter, the existence of this second counter would not change the predictions made by the model.

4. ANALYSES OF PERFORMANCE ON COMPUTATIONAL TASKS

A third way of doing addition is even more sophisticated than the second. According to this model, the counter is set first to whichever of the two numbers is greater; then the incrementing procedure is used for the other number. Thus the problem 3 + 4 is solved by setting the counter to 4 and incrementing 3 times. This saves time because there is always a minimum of counting to be done. Besides being more efficient than the other models, this procedure indicates a higher level of mathematical understanding. The person operating under this model understands that, in terms of the final answer, 3 + 4 is always the same as 4 + 3. It makes no difference in which order they are added. Formally stated, this is the mathematician's *law of commutativity*. The procedure is also a bit more complex in that it requires a decision as to which is the larger number, hence the one from which to begin counting.

If we place these three addition models in their real-time context, each predicts a different duration for task performance. In each case, we assume that the initial setting of the counter takes a fixed amount of time. It makes no difference whether it is being set to 0, or 3, or 7; the time needed is the same. This means that in evaluating the three models, we need to compute only the time it takes for incrementing the counter. Under the first model, task performance would take the time needed to tick off both numbers on the counter (e.g., in the problem 2 + 7, the time needed to count to 9). Under the second model, the counter would be set at 2, and performance would take as long as needed to count out the remaining 7. Under the third model, it would take a very small amount of time to pick out the larger number[3] and set the counter to begin at 7, plus the time needed to count out the remaining 2. Labeling the parts of the equation m, n, and p, such that $m + n = p$, we can tabulate the predicted performance times for the three models as follows:

Model A: Reaction time = the sum of $m + n$.

Model B: Reaction time = n.

Model C: Reaction time = either m or n, whichever is smallest.

Figure 4.4 shows the predictions that would be made for each of the three models. In the figure, the individual problems are designated by two numerals, such that 72 refers to 7 + 2 = □, 45 refers to 4 + 5 = □, and so on. The models differ in the amount of time predicted for each individual addition equation. For example, consider where the problem 4 + 5 appears in the three graphs. Under Model A, 4 + 5 takes nine counts, so it appears next to the predicted latency for nine increments. Under Model B, the counter would initially be set at 4 and five

---
[3]The model assumes the quantity comparison takes a constant amount of time that does not alter the overall reaction time predicted by the model. Actually, there are data suggesting such quantity comparisons are not all equally difficult (Moyer & Landauer, 1967). Deciding which number is smaller seems to be easier (quicker) the farther apart the numbers are (e.g., comparing 7 and 2 is easier than comparing 4 and 5).

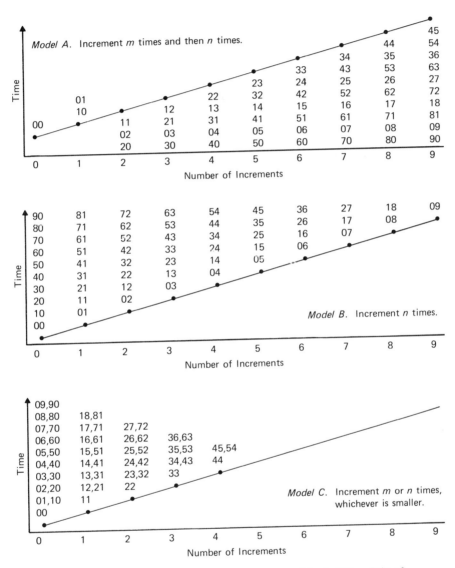

FIG. 4.4 Predicted reaction-time patterns for three models of addition. Pairs of numbers listed above or below dots stand for single-digit addition problems (e.g., 0 + 0, 0 + 1, 1 + 0). Dots indicate the reaction times required to add particular pairs of numbers as hypothesized under the respective models.

78   4. ANALYSES OF PERFORMANCE ON COMPUTATIONAL TASKS

increments then made, so 4 + 5 appears with the latency of number pairs requiring five increments. In Model C the counter would initially be set to 5, so the problem 4 + 5 appears with the equations requiring only four increments.

Suppes, Groen, and others have compared individuals' performances to these models to see which model was best "fit" by the data. The general strategy in these experiments was to present a child each of the addition problems shown in Fig. 4.4 and to measure reaction times (*latencies*) separately for each problem. This was done by having children press a button to answer or by having them speak into a microphone attached to a relay. Either the button or the microphone attachment stopped a clock timer, so that time data were registered along with answers. Each child's pattern of latencies was compared with the predictions of the models under consideration.

Figure 4.5 shows the results of one such study involving first graders (Groen & Parkman, 1972). Averaging all scores for first graders, it compares observed

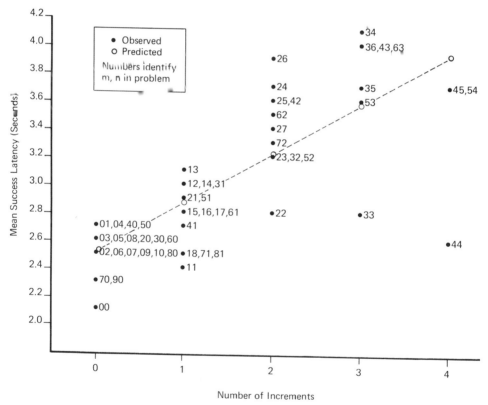

FIG. 4.5 Average reaction times on single-digit addition problems for first-grade children. (From Groen & Parkman, 1972. Copyright 1972 by the American Psychological Association. Reprinted by permission.)

latencies to predicted latencies for Model C. The dots scattered about the graph each indicate the average time required to complete the specific equation designated by the pairs of digits next to each dot. The broken line that slopes upward gives the location (open circles) that the dots would assume if performance conformed perfectly to Model C. The dots more or less cluster about the sloping line, although they seldom match up perfectly. The data are judged to "fit" the prediction because the observed dots *as a whole pattern* do not deviate significantly from the pedicted positions. The degree of fit to various models is tested statistically and the model that accounts for the most variance (i.e., that keeps the dots clustered most closely along the predicted line) is chosen as the one that best describes performance. In this and other experiments, virtually all the first-grade children studied produced data that best fit the predictions of Model C. It is therefore reasonable to conclude that Model C, setting the counter to the larger addend and incrementing by the smaller, describes what most end-of-first graders do.[4]

Transitions in Competence. If a similar kind of experiment is applied to children of two different ages, it is possible to get an idea of how computation strategies change with age and perhaps with instruction. In a study analyzing children's procedures for solving subtraction problems of the form $m - n = r$ (Woods, Resnick, & Groen, 1975), the experimenters compared second and fourth graders. The first step was to hypothesize models for the subtraction process and then to predict latency patterns for each model. Two basic strategies for subtraction were hypothesized: Set the counter to m and count down (decrement) from there n times (Model A); or set the counter to n and count upward (increment) until you reach m (Model B). Depending on the size of the difference between m and n, either Model A or Model B could be the quickest. This suggested a third model: Do *either* the incrementing *or* the decrementing procedure, depending on which will be quickest (Model C). This last was called the "choice model." Under it, $8 - 3 = \square$ would be solved by setting the counter to 8 and *decrementing* three times; $8 - 5 = \square$ would be solved by setting the counter to 5 and *incrementing* three times. The two problems would thus require the same amount of time.

The straight decrementing routine (Model A) is close to what we assume is usually taught in schools: Count out m blocks; then remove n blocks and state how many remain, r. It seems likely that second graders, being relatively close to their first formal experience with subtraction, would follow this model. But somewhere along the line, children probably discover it is sometimes more

---

[4]Note, however, that a certain subset of problems, namely the "doubles" (2 + 2, 3 + 3, etc.), do not come out as predicted. Instead, these problems are all solved very quickly. This probably indicates that the counting model is not applied to these particular problems at all but that their answers are already stored in long-term memory and are recalled directly.

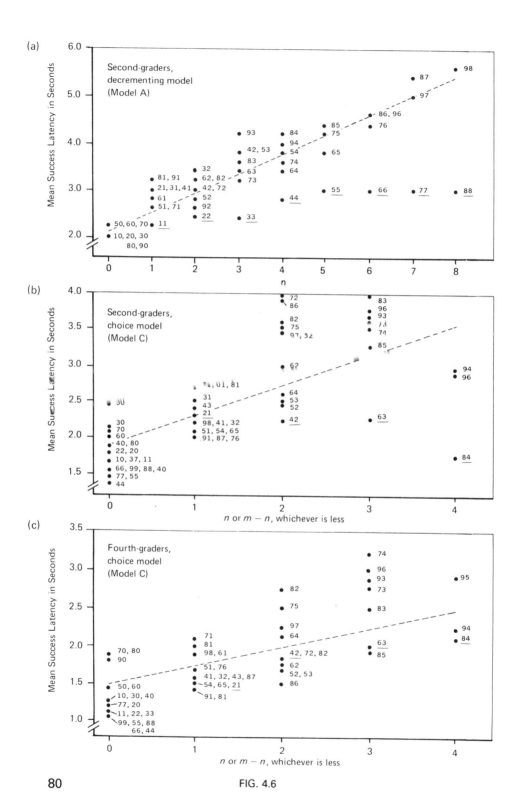

FIG. 4.6

efficient to count up from n to m in order to reach r, and for each particular problem they choose which is the quickest way to find a correct answer. Thus by fourth grade we might expect greater use of the choice model (Model C). This is roughly what Woods et al. (1975) found. One-fifth of the second-grade children had latency data of the pattern predicted by Model A, and the other four-fifths matched the Model C pattern [Fig. 4.6, panels (a) and (b)]. All the fourth graders matched Model C [Fig. 4.6, panel (c)]. It seems reasonable to assume that the change from decrementing to choice is largely the children's own "invention," based on their observations of the relations among numbers during the countless addition and subtraction problems they have computed over months of schooling.

An interesting aspect of these data is that second graders took about 400 msec per increment regardless of which model they used. This is reflected in the slopes of the graphs, which are the same in panels (a) and (b). The slope for fourth graders is shallower, suggesting an incrementing time of only 250 msec. But fourth graders may not actually have been counting each item more quickly. Rather, they may have had the subtraction pairs partially memorized, using the counting routine only when they either could not remember an answer directly or wanted to check their memories. If this were the case, each problem would take less time *on the average*. Thus, a shallower slope would appear, even if every time they did count, they worked at the same 400 msec rate as the second graders. The data from this experiment do not allow us to choose between these two explanations—faster incrementing rate or partial memorization; but the data do confirm that whenever fourth graders do solve by counting, they use Model C.

The addition and subtraction studies described thus far both used subjects who had already been in school for some time. The experimenters guessed, based on common school practice, that these children had been not taught the routines that were observed. To test directly the idea that children actually invent more efficient computation routines than those they are taught, a further study (Groen & Resnick, 1977) was conducted in which the experimenters controlled what the children were originally taught by doing the teaching themselves. To be sure the children could not have been exposed to prior in-school instruction, preschool children were used as subjects.

The children were taught to solve single-digit addition problems of the form $m + n = \Box$ (where m and n ranged from 0 to 5) by using the following algorithm: Count out m blocks, count out n more blocks, and then count the combined set. After a number of practice sessions the blocks were removed and children gave their answers on a push-button timing device. Since practice continued over several weeks, it was possible to note when and if any child switched to the more

---

FIG. 4.6   Plot of reaction times on simple subtraction problems for second and fourth graders. Dots denote average latencies for specific problems. Children start out using the decrementing model, but by fourth grade most have switched to the choice model. (From Woods, Resnick, & Groen, 1975. Copyright 1975 by the American Psychological Association. Reprinted by permission.)

efficient routine of setting a counter to whichever was more, *m* or *n*, and then incrementing by the smaller (Model C in Fig. 4.4). Half of the children studied made such a switch—a clear case of invention since no instruction in the more efficient routine was offered. Thus, children who were taught a certain algorithm changed, after some practice but no additional direct instruction, to a faster and more efficient routine. The improved efficiency was the result, not of faster performance of the taught algorithm but of the fewer steps in the new algorithm, which in turn required a choice or decision on the part of the child.

Fostering Transitions in Competence. Instruction can concern itself with explicitly teaching the routines and procedures that make a person a skilled performer of mathematical tasks. Or, it can seek to create learning situations in which the chances of such transitions occurring are optimized. Transitions in competence that emerge without direct instruction may be more common in children's educational development than we have thought up to now. The empirical task analyses just described reveal differences between the procedures taught to children and the more efficient procedures that characterize skilled performance. The findings suggest that researchers and designers of instruction together should seek to identify similar transitions so that teaching procedures might be specially designed to foster them.

We do not yet know how general the phenomenon of invention is in learning. Research is needed to pinpoint the specific situations in which these transitions occur. But on the basis of available evidence we can predict that teaching routines designed to foster transitions in competence must have certain characteristics. We know that they must make a link between the structure of the subject matter—the mathematical, logical structure, perhaps as described by a rational task analysis—and what we know about actual performance on these tasks from empirical analyses. Rather than being induced directly either from the task structure or from the analysis of skilled performance, we think good teaching routines will have to be *constructed* by the teacher to help children make the transitions in competence that render them skilled performers. We can suggest the following criteria for a teaching routine designed to facilitate inventions:

1. It must display the underlying structure of the subject matter.
2. It must be easy to demonstrate or teach.
3. It must be capable of transformation into an efficient performance routine.

To take an example of this kind of teaching, in the addition study just described, the adding routine taught to the children represented the subject-matter structure clearly. It embodied the "union of sets" definition of addition and was thus a mathematically correct procedure. The routine was also easy to demonstrate and to learn. Four-year-old children, who knew only how to count objects when the experiment began, were performing addition almost perfectly, using blocks, after about a half-hour of practice. But this routine was rather awkward and slow to perform. The 4-year-olds soon recognized that it could be transformed into a more efficient routine, which in turn represented a more sophisticated understanding of the subject matter, since it involved an intuitive appreciation of the law of commutativity.

Another example is found in the subtraction study just described. The children had probably been taught a "partitioning of sets" definition of subtraction. But they invented a more efficient performance routine (the choice model) that was also more sophisticated mathematically, in that it reflected the complementary relationship of addition and subtraction. Thus the teaching routine reflected the subject matter well, was easy to teach, and proved to be transformable into a better performance routine. Why, one may ask, should we not teach the more efficient routines directly? Why rely on children's inventions? In the present cases, the answer lies in the sheer difficulty of explaining the efficient routines to children. Groen and Resnick (1977) tried this informally; they found themselves bogged down in awkward, incomprehensible explanations and found the children bored or frustrated. Yet all the children learned the direct routine taught in the experiment and some, as we have seen, transformed it into the more efficient routine. We can argue, therefore, that the teaching routine Groen and Resnick used was a good one because it put the learners in a position to discover for themselves the better performance routine, although we cannot tell from this study what the process of discovery was.

To summarize, empirical analyses of performance on simple mathematical tasks reveal changes in competence that may be amenable to instruction—not necessarily to direct instruction but perhaps to specially constructed teaching routines that highlight the relationship between subject-matter structure and skilled performance. We will have more to say about the concept of structure in mathematics and its relationship to learning and teaching in the following chapters. But first we need to consider some situations in which it appears that children who do not fully understand the mathematical basis for computational routines use their powers of invention to simplify problems *incorrectly* and thus make consistent errors in their mathematics work.

COMPUTATIONAL STRATEGIES AND SYSTEMATIC ERRORS: PROTOCOL ANALYSIS

Reaction time analyses have been fruitfully applied in the study of relatively simple and straightforward arithmetic tasks, such as judging inequalities ("more than" and "less than") (e.g., Moyer & Landauer, 1967; Restle, 1970; Sekuler, Rubin, & Armstrong, 1971), judging the correctness of simple numerical equations (e.g., Parkman & Groen, 1971), and computing single-digit number facts (e.g., Groen & Parkman, 1972; Suppes & Groen, 1967). But in these tasks performance follows a fairly strict and limited set of rules or algorithms, and they are carried out quickly without much conscious processing. When tasks are more complex—when there are several discrete steps to carry out, when alternate solution strategies could be followed, and when pauses and reconsiderations are usual in the course of solution—reaction times are not a suitable method of study. For analyzing performance on these more complex tasks, psychologists employ several other techniques. In this section we explore the method of protocol analysis in studies of algorithmic computational procedures and of the errors children make in executing those procedures.

The first step in protocol analysis is to take as complete a record as possible of what an individual does in the course of solving a set of problems. This record, the protocol, is then analyzed to reveal regularities of behavior—especially regularities that can be related to theories about how internal information processing proceeds. A "complete" protocol can be very complex indeed. It can make note of physical movements and how long they take—if possible, it even records such fine movements as those of the eyes as they fixate on different parts of a physical display. The protocol records, in order, every action taken by the experimental subject in relation to the problem—solution attempts whether successful or unsuccessful; oral comments made during solution; written notations if there are any. Finally the protocol can include a record of people's thoughts as they work on the problem. For this purpose, subjects are often asked to "think aloud" as they work, saying everything that comes to mind. This verbalized thinking becomes part of the total protocol and is coordinated in a timed record with the physical movements and other solution-related actions.

It is important to note that "thinking aloud" is just one aspect of the total protocol; it has no special status. There was a time in psychology when subjects' own interpretations of their actions were taken as direct evidence of what they actually did. Today's thinking aloud protocols do not assume that subjects "know" what they are doing, only that they can share some of what is conscious in their minds as they work with the experimenter. It is the experimenter's job, then, to use this information, together with more objective observations, to piece together an explanation of the actual problem-solving process.

An informal example of protocol analysis is Lankford's (1972) analysis of the behavior of seventh graders on a variety of typical computational problems where 176 pupils were interviewed individually. The basis for discussion was a test comprising 36 problems in addition, subtraction, multiplication, and division of whole numbers and fractions, and a series of quantity comparisons. In what were characterized as "diagnostic interviews," the experimenter asked the children to read the problems out loud and then compute the answers, verbalizing their thoughts while carrying out the procedures. The resulting data included calculations written on the test sheets, and tape recordings of both student verbalizations and the experimenter's occasional questions designed to clarify the students' thinking. Lankford analyzed the various solution strategies and their resulting

Computational Strategies and Systematic Errors   85

calculations and uncovered some interesting patterns of behavior. The analysis of data supported Lankford's (1972) hypotheses that:

1. Patterns of thinking—computational strategies—which pupils develop in their study of elementary mathematics are highly individualized and often do not follow the orthodox models of textbook and classroom.
2. There are observable differences in the . . . computational strategies of successful computers and unsuccessful ones.
3. Clues for remedial teaching of computational skills may be derived from an examination of the patterns . . . of pupils who are unsuccessful computers [p. 3].

Let us look at a few examples from the data. First, the protocols reveal tremendous variability in the strategies students used to solve the problems. This means there were many deviations from the algorithms taught in the classroom. Some of these unorthodox strategies were successful; others were not. Consider for example the range of strategies adopted by different students in computing multiplication problems:

$$
\begin{array}{ccccccccc}
20 & 19 & 20 & 19 & 19 & 19 & 19 & 19 & 19 \\
\underline{19} & \underline{20} & \underline{19} & \underline{20} & \underline{20} & \underline{20} & \underline{20} & \underline{20} & \underline{20} \\
180 & 00 & 180 & 00 & 3800 & 19 & 218 & 00 & 20 \\
\underline{20} & \underline{38} & \underline{200} & \underline{380} & & \underline{28} & \underline{00} & \underline{218} & \\
380 & 380 & 380 & 380 & & 299 & 218 & 2180 &
\end{array}
$$

Only seven students thought simply "2 × 19 = 38 and add 0" (Lankford, 1972, p. 23). Some of the students' deviant strategies worked to produce correct answers. Others did not. Where wrong answers occurred, examination of individual patterns made it obvious that students' grasp of the number facts was not always at fault. Rather, errors were created by the students' particular computational strategies.

Second, there seemed to be certain patterns in the pupils' strategies. An important and consistent finding was the tendency for many students to use counting in their calculations—even though these students were supposedly well into the stage where "automatic" knowledge of arithmetic combinations is usually expected. Of the 176 seventh-grade students, 93 used counting in adding whole numbers, 63 in subtraction, 63 in multiplication, and 4 in division. Often counting was used to reconstruct unknown from known combinations. For example, to find 7 × 8 a student might say 7 × 7 = 49 and then count up 7 (using fingers!) to reach 56. Operations on fractions resulted in far more errors than operations on whole numbers, and counting was used less for obvious reasons. The data showed considerable confusion in students' algorithms for working with fractions. For example, an important source of error in adding fractions was simply adding numerators and denominators, as follows: $\frac{3}{4} + \frac{5}{2} = \frac{8}{6}$ or $\frac{3}{8} + \frac{7}{8} = \frac{10}{16}$ [Lankford, 1972, p. 30].

4. ANALYSES OF PERFORMANCE ON COMPUTATIONAL TASKS

Lankford selected a group of "good computers" and "poor computers" from the pool of experimental subjects for the purpose of comparing the strategies used by the two groups. Good computers executed the taught algorithms correctly and were able to keep track of the steps as they went along, whereas the poor computers seemed to have trouble remembering the conventional algorithms and the problems to which they apply. Moreover, when poor computers ran into difficulty, they "improvised" procedures, and these often generated errors. Good and poor computers differed in their command of number facts, poor computers more often reconstructing number facts from a small repertoire of known facts, usually "doubles" (8 + 8, 4 × 4, etc.) or products having 5 as a factor. Good computers also seemed more able to sense the wrongness of errors and to go back and check. In long division, they were able to generate estimates of quotients and to discard unlikely candidates quickly, whereas poor computers often tried multiplying several possible quotients by the divisor in an effort to find a suitable quotient.

Probably the most important conclusion to be drawn from Lankford's work is that students' patterns of errors seem to reflect fundamental misunderstandings of procedures rather than random mistakes in carrying out basically correct procedures. According to Lankford, "What appear to be careless errors of poor computers are often supported by a reason—even if faulty. For example, in multiplying 304 by 6 a pupil wrote 1804, seemingly failing to add the carried 2 to the 0 of 6 × 0 = 0. Actually the pupil said '6 times 4 is 24; put down the 4 and carry 2 (written above 0 of 304); 0 times 2 = 0; 6 times 3 = 18' [p. 39]." Where such errors appear in a consistent pattern for any individual, this indicates a faulty understanding of the steps of the procedure. By examining closely the errors made, we obtain a picture of the way that procedure is represented in memory, even if it is represented in an incorrect way.

This view of the source of errors in computation stands in contrast to the one pursued by Suppes and his colleagues in the work described in Chapter 2. Suppes (see Suppes & Morningstar, 1972) tried to explain errors in his drill and practice programs at Stanford by developing a formal mathematical model that could predict the probability of error on any particular problem. For addition problems requiring carrying, the data were best explained by a model that predicted error on the basis of the following factors:

> Probability that any pair-sum will be wrong, i.e., that any particular number fact will be incorrectly recalled (output error).
> Probability of failure to carry.
> Probability of generating an erroneous carry.
> Probability of generating an erroneous carry when the preceding column received a carry (suggesting the influence of problem context upon solution strategy) [Suppes & Morningstar, 1972, pp. 137, 142].

For any particular problem, the probability of error would be determined by summing the above probabilities depending on the specific features of the problem. The idea was to use the probability ratings to order problems according to difficulty, for the purpose of optimizing computer-assisted practice on computation. But Suppes' model assumed that errors were random and thus did little to explain what caused them or to suggest what instruction other than more practice might help to correct them.

There is further confirmation of Lankford's suggestion that errors are systematic rather than random: Ginsburg (1977a, 1977b) and Brown and Burton (1978) have also discovered instances of consistent error-producing algorithms in elementary school children. Brown has begun a program of research specifically geared to mapping out children's procedural knowledge about computation. He describes the misunderstandings or mistaken algorithms as "bugs" in the mental programs that are responsible for executing computations. Take as an example of a "buggy" procedure the following problems as they were solved by one child acknowledged as having trouble with his math. What is the "bug"?

```
    7      9    17     87    365    679    923   27,493
   +8     +5    +8    +93   +574   +794   +481   +1,509
   ──    ──    ──    ──    ───    ───    ───    ──────
   15    14    25    11    819    111    114    28,991
```

Since many of the number pairs are added correctly, one might initially conclude that the wrong answers are due to random error. But a close inspection of the problem solutions reveals a systematic error of procedure that accounts for these specific errors and also predicts how this child would answer other similar problems. Notice the number of "ones" in his problem solutions. The bug in his procedure for column addition is that whenever there is a number to be carried into the next column, he writes down the tens digit (which should be carried) and simply ignores the units digit. Presumably the first three sample problems are solved correctly because they are computed by some other algorithm, such as counting, made possible by their small sum. The same bug shows up in his multiplication, along with another consistent procedural error:

```
     68       734      543
    ×46      ×37     ×206
    ──      ───     ────
     24      792      141
```

In the first problem, he multiplies 6 × 8, correctly finds 48, but notates the 4 and ignores the 8. Then he multiplies the 4 × 6 in the second column, and writes 2 for the 24. The same bugs are seen in the other two problems as well. What at first appears to be very erratic computational behavior is actually strict adherence to an algorithm, albeit an incorrect one. According to Brown and Burton (1978), "A common assumption among teachers is that students do not follow procedures very well and that erratic behavior is the primary cause of a student's inability to perform each step correctly. Our experience has been that students are remarkably competent procedure followers often but that they often follow the *wrong procedures* [p. 157]."

Numerous examples of such consistent patterns of error have led Brown and Burton (1978) to develop models of mental processing which can account for errors as well as correct executions of procedures. In this context they are experimenting with a computer-based training program, BUGGY, that helps teachers identify bugs in individual children's computations. BUGGY trains teachers to interpret a particular child's poor computational performance in terms of the buggy routines that produce it. The rationale underlying BUGGY is that in order to pinpoint errors, one must have a "map" of the procedure showing most of the places the student is likely go wrong. Brown and Burton have simulated these hypothesized mental maps, or "procedural networks," on a computer. The representation of such network models is discussed in greater detail in Chapter 8.

The instructional interest of this work, along with Lankford's (1972), lies in its pointing out that information processing can be systematic even when it leads to incorrect responses or is based upon incorrect information. Consequently, teachers should look for more than errors in number facts to account for poor algorithmic computation; they should attempt to follow the steps of solution with individual children to find where they are going wrong. Lankford's work clearly demonstrates the potential usefulness of diagnostic interviewing for the classroom teacher as well as the psychologist. Brown and Burton foresee a time when schools might have a diagnostic specialist who would work with children having special difficulty in math. This diagnostician would conduct in-depth interviews in conjunction with specific computational tasks to detect possible procedural bugs. The intent would be to gear instruction to specific procedural difficulties of an individual. But no one is suggesting that only the algorithms taught should be considered acceptable. If idiosyncratic strategies are mathematically correct, they should also be acceptable, provided they result in successful problem solutions.

Referring back to our earlier discussion of the place of invention in learning, it can be seen that inventions do not always lead to finding mathematically correct ways of solving problems or performing computations; on the contrary, many very creatively invented algorithms lead to serious computational errors. Are there ways of initially presenting computation routines that will optimize the likelihood of fruitful inventions and minimize the chances of error-producing ones? The question, posed in this form, has received little study, but those who advocate mathematics teaching based on structural principles (see Chapter 5) are operating on the basis of strong intuitions in this direction. In any case, evidence concerning the many forms—both productive and counterproductive—that invention can take implies the need for more individualized attention to children's mathematical performances on the part of teachers. If it is not possible to verify the course of individuals' procedural learning during the teaching sequence, then some form of the diagnostic interview should at least be used with students who are having obvious difficulty with computations.

SOLVING CONTEXT-EMBEDDED PROBLEMS: COMPUTER SIMULATION

Our empirical analyses of mathematical behavior have examined the strategies children use in executing simple and complex computations. Most children routinely carry out these computations both in drill exercises and in tests assessing basic understanding of number concepts and quantity. But with the exception of these exercises, most of the arithmetic we do is in the context of some kind of real problem. It may be a simple context, like figuring out how to divide a cache of candies equally among several children. Or it may be complex, involving a series of multiple comparisons, like figuring out how to optimize returns on stock investments. Whatever the specific context, outside the classroom we rarely engage in calculation alone. Rather, we encounter situations that require that certain kinds of calculations be performed in order to obtain necessary information.

The importance of contextually based arithmetic is widely recognized in educational practice, as witnessed by the predominance of "word" or "story" problems in the curriculum. It is not always easy for the teacher to set up facsimiles of the real world in which actual mathematical problems are encountered. But it is not difficult to present verbally the kinds of problem situations the real world presents in which "applied arithmetic" is required.

A student confronted with a word problem must somehow translate the sentences into a computational problem. This involves reading and interpreting the words, determining what operations are required, and setting up the problem so that known solution procedures can be followed. A variety of linguistic skills—recognizing nouns, adjectives, and verbs and using referential cues in the language—are required. There must be some strategy for identifying what is known and what must be found out. General schemes for interpreting stories, such as those being investigated in today's work on story comprehension (e.g., Anderson, Spiro, & Anderson, 1978), also come into play. In Chapter 2 we saw that the relative difficulty of word problems can be predicted on the basis of their grammatical complexity, number of required operations, the context set by previous problems in the exercise, and so forth. These variables have a bearing on the overall amount of processing that has to be done in order to set up the various problems for calculation. But a more detailed look at the processes of solving word problems is also possible. We look now at one such effort.

A Computer That Solves Word Problems

As we just saw, the purpose of interpreting and analyzing protocols of human mathematical behavior is to construct a theory or model that will account for the main aspects of the subject's performance. To do this, the psychologist must both ignore certain details that are too specific to have general implications for a model *and* fill in or hypothesize some aspects of performance that have not been directly observed. The analysis, then, is a rather artful procedure. More than one interpretation of any given protocol or set of protocols is possible, and this of course raises the question of how to decide between possible interpretations.

One of the tools that psychologists use to help them interpret protocols more rigorously is computer simulation. The basic idea in computer simulation is to prepare a complex model that is capable of solving the class of problems under consideration. If they can get the model to solve these problems, given the same materials and the same instructions one might give a human, they know that the "theory" of problem solving exemplified by the model is at least a possible one; it is capable of solving the problems. The role of the computer is simply to carry the model through each of its (many) steps.

The next question, though, is whether the model describes what *people* actually do. If not, then the model is inadequate as a psychological explanation. Here is where computer simulation and protocol analysis must join. If the model working on a problem produces a record of behavior that matches a human protocol on the same problem, then the model can be accepted as a reasonable one for describing human performance. For example, does the model make errors similar to those made by humans? Does it get "confused" (lose information or misinterpret information) at the same points as humans do? Does it take especially large amounts of time at the same points in the solution process as humans? All of these are indications of a good match between the model of performance, which the computer program represents, and the protocol of actual human behavior. The better this match, the more confident the psychologist can be that the model or theory embedded in the computer program is correct.

Algebra word problems have been studied by Paige and Simon (1966) using computer simulation and protocol analysis methods. They began with a hypothesis about the processing requirements of word problems, a hypothesis expressed as a computer program. In honor of the object of its imitation, the program was called STUDENT. Written by Daniel Bobrow (1968), STUDENT was initially conceived as an "artificial intelligence" program, that is, it was an attempt to build a computer program that could successfully solve algebra problems but not necessarily in the same way that humans did. Rather the aim was to be as logical, rational, and efficient as a *computer program* could be. Although STUDENT was not intended as a model of human information processing, it provided a useful starting point—a formal rational analysis of the task of solving word problems. Paige and Simon took it as such. Through their research with human subjects of various ages and mathematical backgrounds, they set out to see how good STUDENT was as a theoretical model of *human* problem solving. They were looking for ways of modifying the program to match human performances better, so as to build an empirical (not just a rational) description of human processes in this important area of behavior.

Let us have a closer look at STUDENT. STUDENT's talent lies in translating verbal statements into algebraic ones. The processes underlying this translation are what constitute a theory of how word problems are solved.[5] Consider, for example, a fairly typical word problem:

> If the number of customers Tom gets is twice the number of advertisements he runs, and the number of advertisements is 45, what is the number of customers Tom gets?

The first step is to partition the problem into phrases:

> If / the number of customers Tom gets / is / twice / the number of advertisements he runs / and / the number of advertisements he runs / is / 45 /, what is / the number of customers Tom gets / ? /

Then the phrases are translated into algebraic terms:

the number of customers Tom gets:	x
is:	$=$
twice:	2 * (* stands for times)
the number of advertisements he runs:	y
is:	$=$
45:	45
what	?

This direct phrase-by-phrase translation permits the writing of some equations:

$x = 2 \times y$
$y = 45$
$? = x$

These can then be simplified into a single equation:

$x = 2 \times 45$

This example, while true to the spirit of STUDENT, does not really display the complexities of the processing involved. First, STUDENT "knows" some

---

[5]This description of STUDENT is adapted from Roman and Laudato (1974).

rather sophisticated things, such as the fact that it is useful to give arbitrary symbolic names (like x and y) to phrases and that it is important always to use the same symbol for the same phrase. Second, it knows, within limits, how to recognize phrases as equivalent even when they are not identical. Thus, "the number of advertisements Tom gets" and "the number of advertisements" are both assigned the name y. Other mechanisms in the program allow STUDENT to recall certain facts, such as "rate times time equals distance," "three feet equals one yard," or "the plural of child equals children." STUDENT can also recognize special verbs that indicate time sequences, and it has something like a dictionary of terms that imply specific arithmetic operators, so that addition, subtraction, multiplication, division, and squaring can be called for as appropriate.

Comparing the Computer and the Human Problem Solver

STUDENT's processes and knowledge can be viewed initially as analogs of the processes and knowledge humans use. Thus, using STUDENT as a hypothetical model of performance, one can examine human problem-solving behavior for evidence of processes such as identifying phrases, identifying and using knowledge external to the problem, assigning symbolic names, determining when two phrases have the same referent, and so forth. For the purpose of such a comparison, Paige and Simon (1966) presented word problems to individual students and explicitly asked them to try to set up algebraic equations that could be used to solve them. The individuals were asked to "think out loud" as they worked. To make this more likely they were not given pencil and paper but were told they could ask the experimenter to write on a blackboard anything they needed to keep in mind. What was said and written was recorded word for word, and these protocols were then analyzed to see if the processes built into STUDENT could account for what the human subjects were doing. Here is a sample protocol. The problem is read first:

> 'If three more than a certain number is divided by 5 the result is the same as twice the number diminished by 12. What is the number?'

Then the human problem solver begins overt solution activity:

> 'Three more than a certain number' is x plus . . . $x + 3$. Write down '$x + 3$.' 'Divided by 5,' so divide the whole thing by 5. 'The result is' . . . 'is'—an equals. Write '='. 'The same as twice the number,' which would be 2 times x. 'Diminished by,' minus—'minus 12.' That's the completed equation [Paige & Simon, 1966, p. 68].

This protocol clearly shows that at a general level the human subject is proceeding much as STUDENT would. She is breaking the problem into phrases, assigning symbolic names, and thereby producing an equation. Her behavior is typical with respect to the general processes and in that sense confirms STUDENT as a model of word problem solving. At a more detailed level, however, Paige and Simon found evidence for humans' using a number of processes that STUDENT did not have. Their analysis of *human* information processing for this task domain was therefore expanded.

Human performance differed from the computer's in several ways. In the first place, the human problem solvers were much better at recognizing that different phrases referred to the same quantities (e.g., "gas consumption" and "number of gallons of gas used"). STUDENT could recognize the equivalence of different phrases only within very narrow limits. Humans were able to do this much more flexibly. Part of the reason is that humans can use information not explicitly stated in the problem. They have this external information in their long-term memories, and they know when to apply it—information such as the relative value of quarters and dimes; transitivity relations that say quarters are worth more than dimes; and if there are more quarters, then it is impossible for the collection of dimes to be of greater value than a collection of quarters. More recent computer models have built in this kind of knowledge.

Paige and Simon (1966) also noted that some human subjects proceeded not by simply translating the verbal statements into algebraic equations (which was all they were requested to do), but by constructing a physical representation of the problem and then drawing information from that representation. A dramatic example is provided by the following problem:

> A board was sawed into two pieces. One piece was two-thirds as long as the whole board and was exceeded in length by the second piece by 4 feet. How long was the board before it was cut?

A direct translation, such as STUDENT's, might produce the following sequence:

> Let x equal the whole board.
> Then $\frac{2}{3}x$ equals the first section.
> Then $x - \frac{2}{3}x$ equals the second section.
> The second section is 4 feet longer than the first section;
> therefore, $(x - \frac{2}{3}x) - \frac{2}{3}x = 4$ feet.

If one solves the equation derived by direct translation, x turns out to be a negative number. The length of the original board cannot be negative, obviously, but since the experiment required subjects only to set up the equations and not to solve them, not all subjects noticed that the problem was an "impossible" one. Most of those who did notice had not proceeded by direct translation but had instead constructed a physical representation of the situation described. For example:

first section	second section
$\frac{2}{3}x$	$\frac{1}{3}x$

It is clear, simply from inspecting this diagram, that the second section cannot be longer than first.

Paige and Simon's (1966) comparison of the computer model with human behavior thus suggests there is more than one way to go about solving problems. It appears that combining the strategies of physical representation and verbal translation is probably optimal in solving algebra word problems. Implications for instruction seem clear enough: Both procedures should be encouraged and directly taught when necessary. We know of at least one effort to teach children the translation process that turns a word problem into an algebraic equation. This computer-assisted instructional program (Roman & Laudato, 1974) gives practice in representing verbally stated phrases with algebraic symbols and provides increasingly explicit hints to guide the proper translation. Studies using fourth graders in a school setting indicate relatively greater gains not only in word problem-solving skills but also in general mathematics achievement for children who had this kind of practice (Roman, 1975). As far as the strategy of physical representation is concerned, it need not be limited to cases in which an overtly physical problem is presented. Quantities can be represented in terms of number lines, rectangular spaces, or the like. Training in devising and using such representations could be made an integral part of the algebra and prealgebra curriculum, just as it now is in geometry.

Perhaps most important for instruction, this computer simulation work shows that computation cannot be divorced from meaning, especially when problems are context embedded, as they most often are in real-world problem situations. Analysis of protocols of human problem solvers shows that as a model of human problem solving STUDENT is inadequate. The translation into algebraic notation is just one way a person can proceed in solving word problems, but it is the only way built into STUDENT. A model is needed that includes the possibility of different conceptualizations of any problem, which are likely to be differentially effective in helping to identify appropriate computational strategies. This leads us into the psychological question of what constitutes understanding and conceptual knowledge. And it leads into the instructional question of how to teach meaningfully so that knowledge and understanding are sufficiently rich and versatile to handle the complexities of context-embedded problem solving. These are the topics to be taken up in the remainder of this volume. With respect to computer simulation as an aid to analyzing mathematical performance, we describe more recent efforts that are coming closer to describing the complex linguistic processing, conceptual representation, and strategy seeking that are characteristic of human problem solving.

SUMMARY

Our introduction to empirical analysis has focused on several methodologies that uncover the mental processing required in performing mathematical tasks—reaction time studies, protocol analysis, and computer simulation. Reaction time studies have compared fine-grained time measurements on individuals doing specific tasks in order to verify hypothesized performance models. Studies of children's processing during quantification, a skill that appears prerequisite to much mathematical learning, suggest that subitizing is a way of chunking perceptions so as to lighten the processing load on working memory. A series of experiments on addition and subtraction processes reveal that the way we teach children and the way they ultimately perform are not always identical. Instead children often invent new ways of performing arithmetic procedures. At least sometimes these are faster, more efficient, and based on a higher level of mathematical understanding. We suggest that psychological and educational researchers be on the lookout for situations in which such inventions occur and that they help teachers construct special teaching routines to foster these transitions in competence.

The other side of invention is that, unchecked, it may lead to continuing errors in computation and frustration in learning mathematics. Analyses of protocols uncover a variety of strategies, some correct and some incorrect, that children apply to computational problems. The errors children make are particularly instructive because they reveal both poor learning of standard algorithms and a tendency for some children toward systematic deviations from those algorithms. Errors are related to these systematic deviations.

Context-embedded problems, exemplified by word problems, draw upon the processes of language understanding and strategy seeking as well as upon basic computational skills. We have shown how psychologists have used computer simulations in conjunction with detailed protocols of actual human performance in order to refine their understanding of how people solve word problems. Because different representations are likely to be differentially effective for any given problem, we suggest that practice in translating word problems into algebraic equations should be combined with training in producing physical or mental representations, so as to take full advantage of the various processing strategies a problem solver can bring to bear.

REFERENCES

Anderson, R. C., Spiro, R. J., & Anderson, M. C. Schemata as scaffolding for the representation of information in connected discourse. *American Educational Research Journal*, 1978, *15*, 433–440.

Beckwith, M., & Restle, F. Process of enumeration. *Psychological Review*, 1966, *73*(5), 437–444.

Bobrow, D. G. Natural language input for a computer problem-solving system. In M. Minsky (Ed.), *Semantic information processing*. Cambridge, Mass.: MIT Press, 1968.

Brown, J. S., & Burton, R. R. Diagnostic models for procedural bugs in basic mathematical skills. *Cognitive Science*, 1978, *2*, 155–192.

Chi, M. T. H., & Klahr, D. Span and rate of apprehension in children and adults. *Journal of Experimental Child Psychology*, 1975, *19*, 434–439.

Gelman, R., & Gallistel, C. R. *The child's understanding of number*. Cambridge, Mass.: Harvard University Press, 1978.

Ginsburg, H. *Children's arithmetic: The learning process.* New York: D. Van Nostrand Co., 1977. (a)

Ginsburg, H. The psychology of arithmetic thinking. *Journal of Children's Mathematical Behavior,* 1977, *1*(4), 1–89. (b)

Groen, G. J., & Parkman, J. M. A chronometric analysis of simple addition. *Psychological Review,* 1972, *79*(4), 329–343.

Groen, G. J., & Resnick, L. B. Can preschool children invent addition algorithms? *Journal of Educational Psychology,* 1977, *69*(6), 645–652.

Klahr, D., & Wallace, J. G. *Cognitive development: An information-processing view.* Hillsdale, N. J.: Lawrence Erlbaum Associates, 1976.

Lankford, F. G. *Some computational strategies of seventh grade pupils* (Final report, Project No. 2-C-013). HEW/OE National Center for Educational Research and Development and the Center for Advanced Studies, University of Virginia, October 1972.

Moyer, R. S., & Landauer, T. K. Time required for judgments of numerical inequality. *Nature,* 1967, *215,* 1519–1520.

Parkman, J. M., & Groen, G. J. Temporal aspects of simple addition and comparison. *Journal of Experimental Psychology,* 1971, *89*(2), 335–342.

Paige, J. M., & Simon, H. A. Cognitive processes in solving algebra word problems. In B. Kleinmuntz (Ed.), *Problem solving.* New York: Wiley, 1966.

Restle, F. Speed of adding and comparing numbers. *Journal of Experimental Psychology,* 1970, *83*(2), 274–278.

Romano, R. A., & Laudato, N. C. *Computer assisted instruction in word problems: Rationale and design* (LRDC Publication 1974/10). University of Pittsburgh, Learning Research and Development Center, 1974.

Kullman, R. A. *The WORD PROBLEM program: Summative evaluation* (LRDC Publication 1975/23). University of Pittsburgh, Learning Research and Development Center, 1975.

Schaeffer, B., Eggleston, V. H., & Scott, J. L. Number development in young children. *Cognitive Psychology,* 1974, *6*(3), 357–379.

Sekuler, R., Rubin, E., & Armstrong, R. Processing numerical information: A choice time analysis. *Journal of Experimental Psychology,* 1971, *90*(1), 75–85.

Suppes, P., & Groen, G. J. Some counting models for first-grade performance data on simple addition facts. In J. M. Scandura (Ed.), *Research in mathematics education.* Washington, D.C.: National Council of Teachers of Mathematics, 1967.

Suppes, P., & Morningstar, M. *Computer-assisted instruction at Stanford, 1966–1968: Data, models, and evaluation of the arithmetic programs.* New York: Academic Press, 1972.

Woods, S. S., Resnick, L. B., & Groen, G. J. An experimental test of five process models for subtraction. *Journal of Educational Psychology,* 1975, *67*(1), 17–21.

II MATHEMATICS AS CONCEPTUAL UNDERSTANDING AND PROBLEM SOLVING

Throughout the early chapters of this book we have viewed mathematical proficiency essentially as a matter of skilled calculation. There we considered the ways in which drill and practice could be organized to facilitate the acquisition of speed and accuracy in arithmetic computation. We asked how complex abilities might be built up out of simpler ones and how this ordering principle, embodied in learning hierarchies, might be used to enhance the learning of mathematics, defined in those chapters as arithmetic and computation. We examined the performance of children on computational tasks, looking for explanations of the mental processing underlying their procedures.

There can be little doubt that proficiency in the practical computational skills of arithmetic is important. It follows that teachers and schools must devote time and resources to teaching those skills. But there is more to learn about mathematics than skillful computation. Children need to understand certain basic concepts of mathematics—including but not limited to those concepts that underlie the rules and procedures of simple arithmetic. They also need to learn how to apply their conceptual and procedural knowledge flexibly and correctly in mathematical problem solving. Time and resources must therefore go into teaching children the noncomputational aspects of mathematics. How best to teach mathematical concepts and problem-solving skills is far from clear, however, and few people even agree on what should be taught. In Part II of this book we examine the guidance that psychological research and theory can offer to instruction with respect to a conceptual definition of mathematics.

As we turn to conceptual rather than computational criteria of mathematics competence, we are forced to consider bodies of psychological work whose empirical bases and instructional applications are much less specific than those we have described up to now. In fact, we shall be making some fairly large leaps of inference from data that are often only suggestive. This is uncomfortable, and we express some uncertainty ourselves about the implications of particular lines of research and theory for instruction. But bold interpretation seems to be necessary if we are to take seriously the possibility of a psychology of mathematics concerned with concepts and problem-solving skills. In the remaining chapters we show why we believe that the foundations exist for an empirical psychology of understanding and reasoning in mathematics.

Our key to understanding the conceptual bases of mathematics learning and teaching is the notion of structure. This is a term that has many possible definitions, several of which are treated explicitly in the chapters to come. Structure in mathematics can be thought of as the structure of the subject matter itself, that is, the way the body of mathematical knowledge is internally organized and interrelated. The structure of a domain of knowledge is more or less objectively verifiable; expert mathematicians could probably come to some kind of consensus about the basic structure and content of the field. For our discussion we do not attempt to describe the structure of the subject matter. We take it as a "given," although we are aware that debate continues as to which concepts should be emphasized in instructional presentations to children. Our focus is on the psychological analysis of these concepts—how people understand, use, and learn them—and on ways of teaching the complex relations and patterns (structures) that constitute mathematical knowledge.

Of course, any attempt to teach mathematical structures must take into account the intellectual capacities of the learner. This leads us to consider the characteristics of people's thought processes that allow them to apprehend mathematical structures and think mathematically. We look at three definitions of psychological structure, two of which are explicitly developed in the theories of the gestalt movement and Piaget. Gestalt theory (Chapter 6) focuses on the structure of the psychological field, that is, the tendency for perception and thinking to be organized into functional wholes that dominate the objective elements of experience and determine their interrelationships. Piaget's (Chapter 7) interest is in the logical structures of the human mind that determine people's understanding of mathematical events and manipulations. The gradual development of these structures is thought to depend on the learner's active interactions with the environment.

A third psychological definition of structure (Chapter 8), is derived from current information-processing approaches to the study of cognition. We will see that knowledge structures can be hypothesized for particular learners and their performance on mathematical tasks used to verify the content and organization of their mathematical knowledge. We also examine the effects of particular mental representations and strategies upon the utilization of available knowledge structures in problem solving. To place these psychological definitions of structure in an appropriate instructional context, we begin with a chapter (Chapter 5) on recent approaches to teaching children the structures of mathematics.

5
Teaching the Structures of Mathematics

For about two decades beginning in the late 1950s, conceptual approaches to mathematics education made increasing inroads on traditional computational approaches. Special projects[1] were launched both in the United States and internationally for the express purpose of determining how best to teach children the basic concepts and principles that give coherence to the subject matter of mathematics (i.e., the structures of mathematics). The mathematics curriculum was expanded; in schools, young children were exposed to relatively "advanced" concepts like inequalities, the properties of sets, the use of zero as a number, and the principles underlying decimal notation. Educators grappled with the problem of upgrading teacher training to handle the increased demand for expertise in mathematics (*Goals for Mathematical Education,* 1967). New materials appeared, and some old ones were rediscovered, that were especially designed for teaching the mathematical structures underlying computational procedures. Meanwhile, psychological research sought to explain how children come to understand and use complex mathematical concepts.

In this chapter, we look at the beginnings of this period of widespread interest in conceptual approaches to mathematics teaching. Our focus is on the shared interest of mathematicians, psychologists, and educators in extending the range of topics covered in school mathematics and in developing new teaching methods to make mathematics learning "meaningful." Researchers and curriculum developers oriented toward conceptual approaches seem to agree on the importance

---
[1]Among others, the School Mathematics Study Group (SMSG), the Harvard Mathematics Project, Mathematics Education Research Group (MERG) at the University of Pennsylvania, the University of Illinois Arithmetic Project, the Madison Project, and the Minnemast Project.

101

of fostering in children a strong intuitive understanding of the underlying structures of mathematics. We look at several converging lines of thought suggesting that an understanding of mathematical structures is fundamental to meaningful learning. We also describe some structure-oriented teaching methods and materials designed to promote meaningful concept development. We discuss a developmental theory of thinking that suggests how teaching can respond to children's capacity to build mental representations of the subject matter, and present instructional principles developed by a mathematician/educator who used concrete materials to teach mathematical structures to young children. Finally, we pose a number of questions that the structure-oriented approach raises for a psychology of mathematics.

A DECADE OF CURRICULUM REFORM

The problem of making learning meaningful has been noted by some mathematics educators for many years. Even in Thorndike's time, educators such as Brownell were cautioning against drill and practice as a primary teaching technique, because they felt children would view mathematics as a set of unrelated facts and procedures rather than as complex interrelated structures of knowledge. Early attempts to bring meaningfulness into instruction centered around embedding arithmetic skills and concepts in practical exercises that had relevance to everyday life, such as figuring grocery bills or estimating the cost of several yards of material (Trafton, 1975). But rote methods of teaching persisted well into the 1950s, despite the good intentions of educators concerned with meaningful conceptual development.

Then, in the late 1950s and early 1960s, mathematics education came under the impact of several new developments that stimulated interest in the problem of meaningful learning. After the launching of Sputnik and the advent of the Soviet-American "space race," schools came under pressure to produce quickly students whose mathematical sophistication would be equal to the new space age technology. This implied not only teaching more mathematics but also better integrating children's mathematical knowledge. A period of curriculum reevaluation and reform began, centering on mathematics and the sciences. As in earlier reform movements, the need for meaningful learning was judged critical, but this time new criteria for meaningfulness emerged.

From mathematicians came the suggestion that meaningful learning would result if children were taught the mathematical substrate of concepts and skills, that is, the structures of mathematics. They did not expect children to be able to grasp the formal proofs that constitute the knowledge base of mathematics, but they felt children could appreciate intuitively the concepts and relationships that underlie mathematical procedures. In other words, they advocated a conceptual rather than computational approach to mathematics instruction. The meaningfulness of instruction would depend not solely on the relevance of computational skills to real-life tasks but also on the extent to which it mapped onto the internal integrity of the mathematical subject matter.

In American psychology there were new developments at this time as well. The field of "cognitive" psychology was being born, spurred in part by the rediscovery of earlier theorists such as Bartlett (1932), the Gestalt psychologists (Köhler, 1925; Koffka, 1924), and Piaget (1941/1952). From this new field came a renewed interest in the study of human cognitive processes and suggestions for how mathematics instruction might be made meaningful by responding to specific intellectual capabilities of learners.

Why Teach the Structures of Mathematics?

During this period of curriculum reevaluation, two professional conferences took place that were to shape the aspirations of many mathematics educators over the next decade. One conference, held in 1959 at Woods Hole, Massachusetts, gathered psychologists, educators, physical scientists, and mathematicians to consider general principles and propositions about the nature of learning and teaching in mathematics and the sciences. Another conference, involving mostly mathematicians and held in Cambridge, Massachusetts, in 1963, probed the feasibility of vastly extending mathematics education in the schools. It set forth curriculum proposals and principles that directly affected subsequent developments in mathematics education and related psychological experimentation. Although these two conferences each assembled persons from seemingly different areas of interest, and although the expressed purposes of the conferences differed outwardly—Woods Hole scholars scrutinized the *process* of education; at Cambridge, they looked at its *content*—the conclusions arrived at by the two groups were remarkably similar.

Central to the goals of instruction proposed at both conferences was the teaching of mathematical structures. The mathematicians at the Cambridge Conference sought to revamp the mathematics curriculum throughout the schools to the extent that "a student who has worked through the full thirteen years of mathematics in grades K to 12 should have a level of training comparable to three years of top-level college training [of that day] [*Goals for School Mathematics,* 1963, p. 7]." Obviously, this goal would necessitate dramatic changes in the quality and content of mathematics taught in the early grades. One step in the direction of change would be to teach mathematics as a structured discipline. It would be taught so that there was a reason for each statement or procedure and so that the interrelationships among concepts could be demonstrated and the truth of every concept established or invalidated by reference to already proved assumptions. In other words, teaching would emphasize mathematical structures. The Cambridge group said, for example, "we hope to make each student in the early grades truly familiar with the structure of the real number system and the basic ideas of geometry, both analytic and synthetic, ... with part of the global structure of mathematics.... On this firm foundation we believe a very solid mathematical superstructure can be erected [*Goals for School Mathematics,* 1963, p. 8]."

Participants in the Woods Hole Conference, too, were concerned with teaching the subject-matter structure of mathematics, but their reasons for this approach were articulated in terms of the psychological and educational implications of understanding that structure:

> The curriculum of a subject should be determined by the most fundamental understanding that can be achieved of the underlying principles that give structure to that subject. Teaching specific topics or skills without making clear their context in the broader fundamental structure of a field is uneconomical in several deep senses. In the first place, such teaching makes it exceedingly difficult for the student to generalize from what he has learned to what he will encounter later. In the second place, learning that has fallen short of a grasp of general principles has little reward in terms of intellectual excitement.... Third, knowledge one has acquired without sufficient structure to tie it together is knowledge that is likely to be forgotten [Bruner, 1960, pp. 31–32]

In other words, if teaching could help learners achieve a fundamental understanding of the structure of mathematics by presenting the reasons that underlie mathematical operations and making clear the concepts that link one operation to another, then those learners would ultimately be better able to retain their new knowledge in memory, to generalize their understandings to a wide range of phenomena, and to transfer their specific learnings to novel situations and tasks.

Evident in the proceedings of both conferences was the shared conviction that mathematics is an exciting intellectual achievement and that this excitement could and should be conveyed to children, even very young school children. This would necessitate introducing so-called "advanced" topics into the elementary school but in forms suitable to the capabilities and prior understanding of the learners involved. Suiting the instruction to young learners would not require abandoning the basic structures of mathematics. Rather, it would mean presenting partial or incomplete structures in ways that promote intuitive understanding of the interrelationships among their various parts, such that later learning would serve to "fill out" or complete the structures. The vehicle for this kind of education would be a spiral curriculum, in which topics would be taken up again and again, each treatment being somewhat less intuitive and more formalized than the last and demonstrating relationship to an increasingly wider set of mathematical concepts.

Implicit in this approach to mathematics teaching was a great respect for the intellectual capacity of the young child. Psychologists, educators, and mathematicians alike seemed confident that young children could grasp rather complex mathematical topics, provided the topics were presented in forms appropriate to the children's level of intellectual development. A particularly strong concern was that the drill curricula that still prevailed in schools robbed children of the chance for intellectual excitement and discovery. Convinced also that education had to be responsive to the social and emotional needs of children, they saw drill as tedious and destructive of children's motivation. Their approach to instruction, then, implied moving away from and perhaps even eliminating drill from the mathematics curriculum. Instead, teaching methods were to be used that allowed children to discover certain generalizations and principles for themselves, thus permitting them to enjoy learning and to participate in some of the creative processes that mathematicians have enjoyed through the centuries.

In summary, the long-debated issue of how to make learning meaningful was taken up again by mathematicians, psychologists, and educators in the 1960s as part of a large-scale curriculum reform movement. Drill and practice was to be replaced by learning with understanding, the criterion for which appeared to be (1) teaching that emphasized the structures underlying mathematical procedures and concepts; and (2) teaching that responded to the rich intellectual capacities of the child.

Some Examples of Structure-Oriented Teaching

Just what are the structures of mathematics? Most lay persons have never considered mathematics as anything more than a collection of procedures for solving computations. But any mathematician knows that mathematics is a unified system of concepts and operations that explain certain patterns and relationships that exist in the world. Along with concepts and operations go more or less abstract statements of patterns and relationships, expressed as axioms or rules in mathematical formulas, that give meaning to those patterns in relation to others. And, there is a body of procedures for manipulating mathematical concepts and patterns in orderly and precise ways. Often, these patterns and procedures are discovered "accidentally" or sensed intuitively before they are set down in formal mathematical proofs; both intuition and formal presentation are respectable mathematical activities.

What we know as distinct subject areas (e.g., arithmetic, algebra, geometry, set theory, function theory, calculus) are all components of a larger mathematical system whose parts are intricately dovetailed. Mathematics is not a static body of knowledge. Indeed mathematics, as we know it, has evolved over thousands of years through the cumulative efforts of many individual mathematicians. Its history may be viewed as an unfolding of patterns and relationships and a quest for increasingly accurate symbolic explanations and interpretations for these phenomena.

To understand the structures of mathematics, then, implies grasping both the interrelations among concepts and operations and the rules by which they may be manipulated and reorganized to discover new patterns and properties. Most

106   5. TEACHING THE STRUCTURES OF MATHEMATICS

adults and, up until recent decades, most school children have rarely seen these structural aspects of mathematics (except perhaps in geometry), because school teaching introduced only bits and pieces of the field and rarely related them to the evolving structure of mathematics as a whole.

To give precise definition to all the structures of mathematics would require the expertise of a mathematician and is outside the scope of this volume. But we can give a flavor of the kinds of mathematical structures that children might learn by means of several examples of structure-oriented mathematics teaching. Consider the following sample instructional sequences:

> The teacher displays sixteen sticks or counters and gives Emily the following instructions. "Make a group of five sticks. Are there enough left over to make another group of five?" (Yes) "Make as many groups of five as you can. How many groups of five are there?" (Three) "How many sticks are left over?" (One) "Are there enough sticks left over to make another group of five?" (No) "Okay, now group these counters by sixes. How many sixes are there?" etc. [from Payne & Rathmell, 1975, p. 139].

> The teacher gives Tommy two balance scales, some marbles, and cubes of various sizes. After experimenting with the scales for a while, Tommy finds that the largest cube weighs the same as eight marbles. On the second scale, the largest cube and one small cube together are balanced on the other side by ten marbles. The teacher asks Tommy to guess the weight of the small cube on the basis of this evidence. [from Rasmussen, Hightower, & Rasmussen, 1964, p. q-2].

In the first example, the teacher is giving a lesson on grouping objects into sets and naming the sets. This is thought to prepare the child for numeration and base representation concepts. The exercises require the kinds of exchanges that would be involved in operations in base 5 and base 6. In the second example, Tommy is actually solving simultaneous equations, although he has no idea that his scales and marbles are related to algebraic symbols. Both examples involve the use of concrete objects that children manipulate to find answers to the teacher's questions. They are attempts to convey complex mathematical concepts—numeration, base representation, simultaneous equations—in simple terms so that learning is accomplished with understanding and meaning.

Let us examine in somewhat more detail a possible sequence of instruction designed to reveal the mathematical structures that underlie a typical algorithmic computation—addition with carrying. There are several ways carrying might be taught. For example, a teacher could demonstrate to the child that if the digits in the ones column summed to more than 10, he or she could "carry the 10" by writing a little numeral 1 (or thinking it) over the tens column and adding it in with the rest of the numbers in the tens column. Then the teacher might give the child many sample problems to work in the manner demonstrated. This is a perfectly acceptable algorithm, and one would expect the child eventually to be able to use it accurately and quickly without having to think through the rationale for every step. Many psychologists and mathematicians would argue, however, that such an algorithm should be learned with "meaning," that is, in the context of an understanding of the place-value concept and its structural relation to base systems of notation. How could a teacher present this set of concepts and the procedure of carrying and still remain faithful to the underlying mathematical structures?

Many math textbooks today (e.g., Dilley, Rucker, & Jackson, 1975) introduce place-value notions using pictorial representations of groupings in the base 10 (decimal) system. A teacher who wanted to be sure her pupils understood the principles underlying positional notation would have them work with bundles of sticks or straws or with specially constructed blocks that represent units (or ones), tens, hundreds, etc. As part of an instructional session the teacher might ask them to gather bundles of straws representing, say, 2 hundreds, 4 tens, and 6 ones, and the children would arrange a display as shown in Fig. 5.1a. After having the children arrange groups of straws many times, the teacher would begin to label the groupings with numerals but would still keep the ones, tens, and hundreds separate, as in Fig. 5.1b. Still later, the teacher would show the children how to use "standard notation" and how to say the numbers (that the bundles of straws stand for) in the conventional way (Fig. 5.1c).

As Fig. 5.1 demonstrates, this teaching sequence follows a progression from concrete to increasingly symbolic representations and tries to give the children an

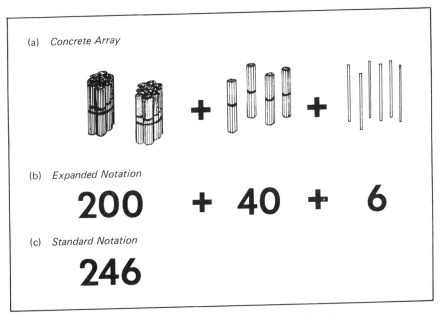

FIG. 5.1 Teaching positional notation using bundles of straws.

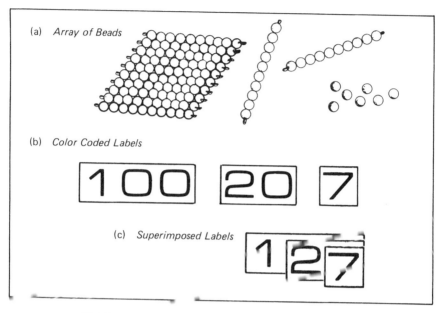

FIG. 5.2 Montessori materials for teaching positional notation.

intuitive understanding of the mathematical realities that standard notation is designed to represent. The point of such a presentation is to get children used to grouping things, not necessarily in multiples of 10 but also in multiples of other numbers, as in our earlier instructional example. This is because understanding grouping notions in general has the potential for making it easier to understand different base systems later on. The point is not universally accepted, however. Some argue that the different bases are merely distractors from the main job of learning the decimal system and that instruction should not admit such confusion.[2]

The Montessori math materials (see Montessori, 1917/1964) represent an attempt to teach place value concretely through the systematic use of color coding and a carefully planned sequence of manipulative materials. The manipulatives consist of various configurations of small beads that embody groupings of units, tens, and hundreds (or groupings in other base systems). In making the transition from representing groups of units, tens, and hundreds through arrays of beads (Fig. 5.2a) to representing them by numerals, the children work with label cards that are color coded to highlight the place value of numerals. Units are notated in green, tens in blue, and hundreds in red. (The same colors

---

[2]Base 10 is often used first because it relates easily to patterns common in daily life, such as our monetary system and metric measurement. The decimal system is actually the culmination of many centuries of mathematical discovery and thinking.

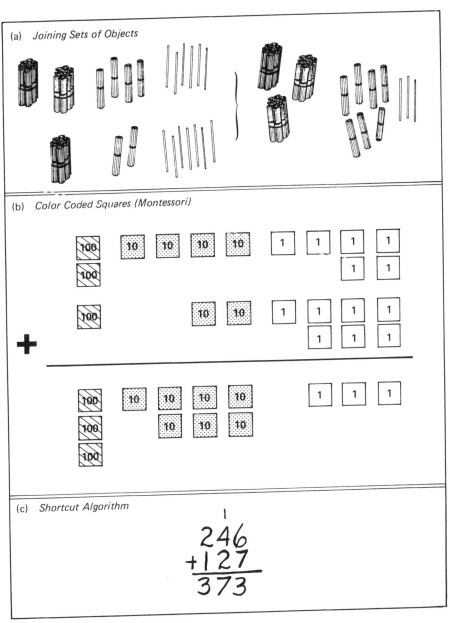

FIG. 5.3 Addition with carrying: Examples of teaching presentations.

are later used for the thousands, ten thousands, and hundred thousands, etc., to highlight the pattern of repetition in the decimal system.) The color-coded labels for arrays of groupings are constructed such that they can be superimposed to produce standard notation (Fig. 5.2c). Thus both sight and touch are involved in teaching the basis for the notational system.

In both of our examples so far, place-value notions have been integral to the presentation of the notational system. If one were not interested in teaching the mathematical structure underlying the notation, then one might merely teach children to read numbers by rote. Structure-oriented mathematicians and psychologists believe, however, that having a chance to "see" what standard notation means by grouping and manipulating objects and actually putting the numbers together on their own helps children learn concepts of numeration and place value in a mathematically honest way.

The benefits of such a structural presentation are supposed to become apparent when instruction proceeds to the algorithm of interest, addition with carrying. A child can set up, or can be shown, a simple column addition problem using straws, sticks, beads, or blocks such as were used to teach numeration (see Fig. 5.3a). When the two sets are joined it is easy to see that there are more than 10 "ones," so that 10 of them could be bundled into a "ten," to be added with the other tens. The Montessori materials provide small wooden squares, each with either a 1, 10, or 100 imprinted on it and colored green, blue, and red, respectively. A written addition problem can be physically performed, and the necessary exchanges of ones for tens made (10 green ones are traded for 1 blue ten) using these little squares (Fig. 5.3b). Only after each child has had many opportunities to execute this sort of exchange physically and to represent the addition operation with appropriate notation does the structure-oriented teacher introduce the shortcut-carrying algorithm (Fig. 5.3c) that we described at the beginning of this discussion. The point of waiting to introduce the algorithm is to avoid using a symbolic representation for a procedure until what it represents is thoroughly understood. One might argue, however, that it is more effective to present the algorithm first and then to introduce the underlying principles once the child is familiar with the steps of the algorithm. This is the sort of question that research has not yet answered definitively.

BRUNER AND THE COGNITIVE REPRESENTATION OF MATHEMATICAL CONCEPTS

The kind of teaching we have been describing would probably meet with a mathematician's approval, on the grounds that it faithfully reflects the underlying mathematical structure of the carrying algorithm. But how well do these teaching methods respond to the specific intellectual capacities of children? If our interest is in building an understanding of mathematical structures in young students, we must do more than simply point out those structures. We must also determine what cognitive capabilities children bring to mathematics learning and how teaching episodes that present those structures interact with children's capabilities. In other words, we must have a theory of intellectual functioning by which to evaluate the likelihood that specific instructional presentations will build the proper understanding. In the remaining chapters we explore several such psychological theories of cognitive functioning and their relevance to the teaching of mathematics.

It is not clear to what extent the curriculum reformers of the 1960s actually drew upon psychology for a theory of intellectual functioning to guide their development efforts. However, one psychologist whose name is strongly associated with the movement is Jerome Bruner. In his own work, Bruner combined the concerns of experimental psychology with those of classroom practice, and his classroom experiments dealt chiefly with mathematics learning. Like many structure-oriented educators who were attempting to develop elegant mathematics teaching and to demonstrate children's capacity to understand sophisticated mathematical concepts, Bruner worked closely with individual children in experimental teaching situations. A strong advocate of close working relations among psychologists, educators, and mathematicians, Bruner collaborated in these classroom experiments with Z. P. Dienes, a mathematics educator whose work we describe later. We present aspects of Bruner's theory of conceptual development as the first of several theoretical treatments of cognitive capabilities as they interact with instructional presentations. Bruner's work also serves as an introduction to the concept of cognitive representation, a topic that recurs frequently as we explore other theories of cognitive functioning in later chapters.

Bruner was one of a growing number of American psychologists who in the years after World War II experienced a revival of interest in human *cognitive processes*—"the means whereby organisms achieve, retain, and transform information [Bruner, Goodnow, & Austin, 1956]." The study of these processes had been overshadowed by the behaviorist orientation of experimental psychology for several decades. Bruner, along with a number of colleagues and students, launched an extensive program of laboratory studies on the cognitive processes involved in thinking and learning (Bruner et al., 1956; Bruner, Olver, & Greenfield, 1966). One focus of that work was concept development. Bruner and his colleagues conducted experiments with adults in which they examined the strategies people employed in the complex process of sorting and classifying—deciding what is and is not relevant to the new idea being shaped—that constitutes concept development. Against this background of laboratory experimentation with adults, Bruner began to examine the cognitive processes of children and became especially interested in how children mentally represented the concepts and ideas they were learning.

Piaget (see Chapter 7) had suggested that development involved successive restructurings of facts and relations, which resulted from children's interactions with and active manipulation of their environment. Building in part upon Piaget's developmental notions, Bruner focused on just how the results of such interactive episodes were represented in the child's mind. In Bruner's words:

> If we are to benefit from contact with recurrent regularities in the environment, we must represent them in some manner. To dismiss this problem as "mere memory" is to misunderstand it. For the most important thing about memory is not storage of past experience, but rather the retrieval of what is relevant in some usable form. This depends upon how past experience is coded and processed so that it may indeed be relevant and usable in the present when needed. The end product of such a system of coding and processing is what we may speak of as a representation [Bruner, 1964a, p. 2].

As we see in later chapters, the concept of cognitive representation has become increasingly important over the years as psychologists attempt to define the processing requirements of learning and problem solving. Of interest for the present discussion is the relevance of different *modes* of representation for the design of math materials for instruction.

Bruner (1964) describes three modes of representation: enactive, iconic, and symbolic. By *enactive* representation is meant "a mode of representing past events through appropriate motor response [p. 2]." This mode is thought to be the only way infants can remember things during what Piaget has called the sensorimotor stage, exemplified by the baby who rattles a fist after having dropped the rattle, indicating it remembers the object in connection with the action performed upon it. Adults too, may represent certain complex motor acts enactively. For example, our muscles do the remembering when we get on a bicycle we haven't ridden for years. This mode may be what we are seeing in children who figure addition problems by tapping their fingers against their chin, or on the tabletop, in an obvious counting motion. Counting for these children may still be represented as a motor act, the same one they started with when they learned to count blocks by tapping each one in succession.

The second mode of representation, *iconic,* takes us a step away from the concrete and physical to the realm of mental imagery. According to Bruner, iconic representation is what happens when the child "pictures" an operation or manipulation as a way of not only remembering the act but also recreating it mentally when necessary. Such mental pictures do not include every detail of what happened but summarize events by representing only their important characteristics. To take an adult example, a person giving directions to a stranger might picture him or herself driving through the streets to the desired destination and might mention the drugstore on the corner before the left-hand turn but not every single house, tree, and fire hydrant along the way. Similarly, a little boy learning seriation might store as pictures his experiences with arranging blocks in order of size, so that future instructions to seriate are understood with reference to images of what he did previously.

Symbolic representation, Bruner's third way of capturing experience in memory, is made possible largely by the advent of language competence. A symbol is a word or mark that stands for something but in no way resembles that thing. It is completely abstract. For example, the numeral 8 does not look at all like an actual array of objects having that number property and neither does the word *eight*. Symbols are invented by people to refer to certain objects, events, and ideas, and their meanings are shared largely because people have agreed to share them. When children begin to write their mathematical operations (using numerals; simple formats such as equations and columns; and the operational signs +, −, and =) this is the beginning of symbolic representation, as is their ability to "read" these mathematical notations. They soon learn to think about their performances in terms of the same symbols, which opens up the new possibility of abstract thinking.

Enactive, iconic, and symbolic modes of representation are related developmentally, according to Bruner. They develop in that order, each mode depending on the one that preceded it and requiring a great deal of practice before the transition to the next mode can occur. This formulation of modes of representation amounts, in Bruner's writings, to a stage theory of the development of intellect. In many ways it is similar to Piaget's theory, having been inspired by the Genevan work, but the two theorists' efforts have received different interpretations in the classroom. As we see in Chapter 7, some interpreters of Piaget emphasize the need to wait until a child is "ready" before trying to teach concepts that depend on the child's possessing concrete operations, formal operations, or the like. Bruner, who has concerned himself more directly with classroom applications, has stated the opposite case: "Any idea or problem or body of knowledge can be presented in a form simple enough so that any particular learner can understand it in a recognizable form [Bruner, 1966, p. 44]." In other words, he felt there were ways to present complicated concepts such that children of any age would understand them at a level appropriate to their intellectual capacities and experience. This confidence in children's abilities was reinforced by his efforts to teach elementary-level children such topics as quadratic equations and the properties of mathematical groups (see Bruner, 1964, 1966).

Extending his developmental work into prescriptions for classroom instruction, Bruner argued that if intellect developed in the order enactive–iconic–symbolic, then it made sense to teach new concepts in that order. This implied that concept development followed a course that paralleled the general theory of intellectual development. For instruction, then, the key seemed to be to present concepts in ways that would respond directly to the hypothesized modes of representation. The ways in which humans mentally represented acts, objects, and ideas could be translated into ways of presenting concepts in the classroom. And even though some students might be quite "ready" for a purely symbolic presentation, it seemed wise, nevertheless, to present at least the iconic mode as well, so that learners would have mental images to fall back on in case their symbolic manipulations failed.

The following passage presents Bruner's instructional notions in the context of a specific mathematical example (see also Fig. 5.4):

> The distinction [between enactive, iconic, and symbolic representation] can most conveniently be made concretely in terms of a balance beam.... A quite young

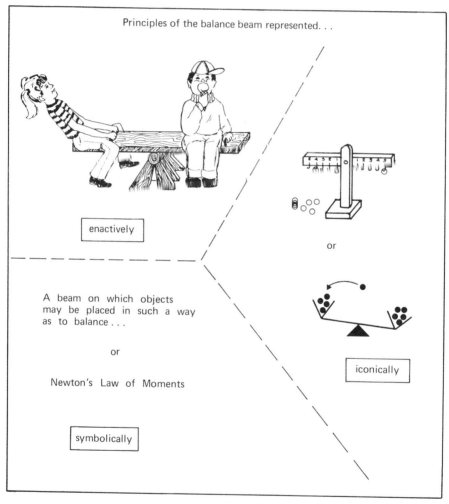

FIG. 5.4 Three ways of representing one mathematical idea in thought. Math concepts may be taught using materials that appeal to particular modes of representation, depending on the developmental level (or level of concept development) of the child. (Adapted from Bruner, 1960.)

child can plainly act on the basis of the "principles" of a balance beam, and indicates that he can do so by being able to handle himself on a see-saw. He knows that to get his side to go down farther he has to move out farther from the center. A somewhat older child can represent the balance beam to himself either by a model on which rings can be hung and balanced or by a drawing. The "image" of the balance beam can be varyingly refined, with fewer and fewer irrelevant details present, as in the typical diagrams in an introductory textbook in physics. Finally, a balance beam can be described in ordinary English, without diagrammatic aids, or it can be even better described mathematically by reference to Newton's Law of Moments in inertial physics. Needless to say, actions, pictures, and symbols vary in difficulty and utility for people of different ages, different backgrounds, different styles. Moreover, a problem in the law would be hard to diagram; one in geography lends itself to imagery. Many subjects, such as mathematics, have alternative modes of representation [Bruner, 1966, p. 45].

Note that our earlier presentation on how to teach the positional notation principle underlying the carrying algorithm also followed a progression corresponding to the sequence of Bruner's modes of representation. Manipulation of straws, blocks, and beads allowed for enactive representation of number concepts. The same materials could be remembered iconically, and the color coding of place value served to enrich mental imagery. The number symbols of standard notation and the symbolic manipulation involved in the carrying algorithm came late in the sequence and built on the experiences that preceded it.

Bruner's theory of the sequence of concept development and his instructional example based on that sequence raise several important questions concerning the nature of cognitive representations. First, should we be satisfied that the three modes—enactive, iconic, and symbolic—exhaust the possibilities for modes of representation? Might there be other ways that concepts and procedures are represented during mental work or might these categories themselves require more elaboration? Second, should we assume, because enactive, iconic, and symbolic representations seem to follow a sequence in children's development, that any one mode is to be preferred over another as a goal of instruction? Is it the case that a symbolic representation is more "advanced" than an iconic representation? It seems more productive to view the various modes as all being useful depending on the requirements of a particular situation. We also need to ask how representations change with developing expertise and how transitions take place from one mode of representation to another. In any case, can we prove experimentally what mode of representation a child is currently using, and can a teacher use this information to tailor instruction to the child's current level? Finally, is the notion of cognitive representation really useful in defining more precisely the processing requirements of mathematical tasks?

These are difficult questions that demand further research, and they should be borne in mind as we explore other approaches to cognitive processing. As we see in further chapters, the theories and experiments of gestalt, Piagetian, and

DIENES' MULTIPLE EMBODIMENTS AND THE SEQUENCE OF INSTRUCTION

The problem of designing meaningful instruction—instruction that takes into account both the structures of mathematics and the cognitive capabilities of the learner—is approached from a mathematics educator's perspective in the work of Zoltan P. Dienes. Dienes devoted a career to designing materials for teaching mathematics and conducting experiments to clarify certain aspects of mathematical concept acquisition. Drawing heavily on Piagetian theory and having worked closely with Bruner on an experimental mathematics project at Harvard, Dienes spoke for the importance of incorporating the research findings of psychology into mathematics instruction. His work stands as one proposal for combining psychological and mathematical principles in structure-based teaching.

The hallmark of Dienes' approach to mathematics instruction is the use of concrete materials and games in carefully structured learning sequences. He is certainly not the first educator to suggest the use of concrete materials, nor have his particular materials been shown empirically to be more productive of learning in comparison to others available. We use Dienes as an example because he has written extensively on the principles underlying the use of concrete materials (see Dienes, 1960, 1963, 1967; Dienes & Golding, 1971). The parallels between Dienes' instructional sequence and Bruner's modes of representation should become apparent as we describe Dienes' principles of instruction. But first, we discuss the relevance of concrete materials to young children's mathematics learning.

Manipulative Materials

Dienes believes that children are by nature fundamentally constructivist rather than analytic. They piece together (i.e., construct) a picture of reality from the experiences they have with objects in the world. This process depends on a great deal of active exploration, as Piaget has emphasized (see Chapter 7). Because mathematical patterns and relationships are not obvious in children's everyday environments, Dienes proposes the creation of teaching materials that embody these structures and bring them within the realm of concrete experience.

As we pointed out earlier, an abundance of specially designed math materials have been produced (or rediscovered) and experimented with in recent decades. These materials have certain features that make them particularly useful in structure-oriented teaching. First, they are free of distractors; that is, they are not used for other purposes in daily life but are clearly designated as objects to facilitate the learning of mathematics. Second, the materials embody mathematical structures without necessarily being tied to symbolic notational systems. They embody qualitative as well as quantitative characteristics of mathematics. Using them, one can become familiar with abstract concepts—the associative, commutative, and distributive laws; the principles of logic; the properties of symmetry; and the number operations—without ever thinking about numbers or writing any symbols. Ultimately, of course, symbols need to be attached to the concepts, but this is deferred until after a series of experiences with the materials.

Dienes has designed his own set of mathematics materials, called *multibase arithmetic blocks* (MAB) or simply *Dienes blocks*. These have gained wide use in mathematics education and research, and modern textbooks often show them in diagrams depicting concrete manipulations in the base 10. The MAB are sets of wooden blocks, each embodying a different base system. Every set has blocks of several sizes that are segmented to show how many units are contained in each. Fig. 5.5 depicts three sets of MAB, one base 2 set and two varieties of base 3 materials. Taking the base 2 blocks as our example, we see there are single cubes called *units* (these are the same size—1 cubic centimeter—in every base set); rods of 2 units, called *longs;* square pieces of area 2×2, called *flats;* and larger cubes with dimensions $2 \times 2 \times 2$, called *blocks*. From the *long* onward, each piece, when multiplied by itself, yields a new physical representation corresponding to the next "power." Were we to multiply a 2 long by 2, we would find a flat measuring 2×2 or 2^2.[3] Blocks, flats, and longs may be constructed from combinations of smaller pieces (e.g., a flat in base 2 could be built of 1 long and 2 units).

A variety of learning experiences are possible using the MAB because they can embody several kinds of mathematical structures. Which structure is highlighted depends on the planned sequence of exercises. For example, with the help of Dienes blocks, children should be able to grasp the notion of exchange underlying the carrying algorithm. If they perform arithmetic operations in different base systems, they may learn that the exchange procedure is the same no matter what base is used. The bottom half of Fig. 5.5 gives an example of addition in base 3. If children unite the two sets shown, John's set and Mary's set, they end up with 3 blocks, 3 flats, 4 longs, and 4 units. With a few exchanges (3 units for a long; 3 longs for a flat, etc.) these can be converted into 1 long block, 1 block, 1 flat, 2 longs, and 1 unit (adapted from Dienes, 1966, p. 19).

In experiments conducted jointly with Bruner (1964b; Dienes, 1963) Dienes used the MAB to embody another mathematical structure, the factoring principle underlying quadratic equations. Children were given flats, longs, and units and

---

[3]We run into problems when it comes to embodying values above about the third power, because, strictly speaking, there should be some unique shape to characterize each of the higher powers. Because representation in a fourth or higher dimension is impossible with concrete materials, the fourth power is generally represented as "long block," and the pattern of long-flat-block is simply repeated as the powers increase.

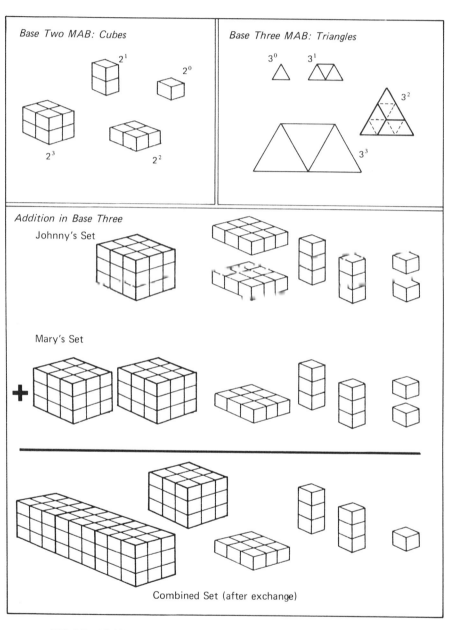

FIG. 5.5 Multibase arithmetic blocks (MAB) created by Z P. Dienes. (Adapted from Dienes, 1966, and Dienes & Golding, 1971.)

were asked whether they could build increasingly larger squares using the materials. The children discovered various ways to accomplish this task. Two examples are depicted in Fig. 5.6. The experimenters later guided the children into devising notations for representing their constructions and employing the same notations with analogous operations performed on a balance beam (another embodiment of the factoring principle).

Attribute blocks are another example of teaching materials designed to embody mathematical structures concretely (Dienes & Golding, 1971). Attribute blocks are wooden or plastic triangles, squares, circles, and hexagons (some sets include other shapes) that vary in size, thickness, and color, such that every piece has a unique set of attributes. In a set of 48 pieces, for example, 24 are thick, and 24 are thin; 24 are large and 24 are small; so that 12 are both large and thick and 12 are both small and thick, etc. Attribute blocks can display principles of classification, set theory, and logic. To discover the properties of sets or principles of classification, a child might be asked to select all the triangles. Within that set, the pieces will vary in size, thickness, and color, but they will have in common the attribute of shape. Simple logic games are also possible with these materials. The teacher, or a child, can keep a set of attributes in mind (e.g., all the blocks that are either thin *or* round), and the other children take turns selecting a candidate block and receiving feedback (a yes–no answer) on whether or not it is a member of the set. Plastic hoops are available so that giant Venn

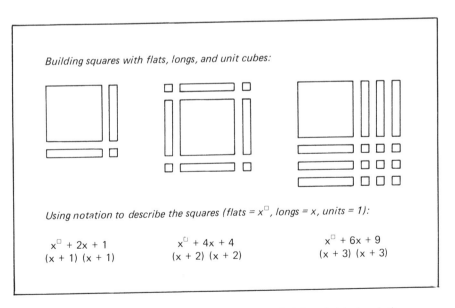

FIG. 5.6 Concrete embodiment of the factoring principle underlying quadratic equations. (Adapted from Bruner, 1966.)

diagrams can be laid out depicting the union of sets and the associated logical principles of disjunction, conjunction, and negation (e.g., pick out all the not-red blocks). Again, symbolic notational schemes can be devised eventually to stand for the principles embodied in the attribute blocks, but they are not needed for children to achieve an intuitive feel for the underlying mathematical or logical structures.

Cuisenaire rods—cubes and unsegmented sticks of various lengths—are an example of materials explicitly designed to embody arithmetic concepts without reference to numeration. They are color coded such that within color "families" certain mathematical relationships are present. Color signals the patterns and relationships to be discovered without involving numbers or other symbols. There is debate over whether it is useful to substitute color relationships for numerical relationships, since the children seem to have to learn first one code and then another. Nevertheless, of the many concept-oriented mathematics materials available, Cuisenaire rods have been among the most readily adopted by classroom teachers. Acknowledging their usefulness for teaching fractions and ratio, Dienes and others nevertheless have expressed some reservations about their exclusive use in the classroom. Dienes was concerned with the possibility that children's learning might become bound to one set of materials, and that this might interfere with the process of abstracting the desired concept, a process to which we turn in the next section. However, the greatest proponent of Cuisenaire rods (Gattegno, 1963) seems to feel the rods are uniquely capable of solving the problems associated with teaching mathematics to children, both because the rods embody the core relationships and structures of mathematics and because they stimulate intuition and inquiry.

The Learning Cycle

Out of Dienes' extensive classroom experimentation has come a scheme for teaching complex concepts to young children, the "learning cycle" (see Dienes & Golding, 1971). Mathematical concept development, according to Dienes, can best be achieved through a series of cyclic patterns, each involving a sequence of learning activities ranging from the concrete to the symbolic. The learning cycle is a planned interaction between one segment of a structured body of knowledge and an active learner, conducted through the medium of specially designed math materials. The reader will note the similarity of this concept to the spiral curriculum proposed by the curriculum reformers of the 1960s.

In each learning cycle, the earliest phase of concept development begins with free play. Children manipulate math materials in unstructured ways, gaining a sense of their size, weight, texture, and color, and discovering the ways they can be used in imaginative constructions. This kind of activity is not purely random, however, for the children are by nature alert to patterns and regularities in the environment. The free-play phase of the learning cycle should not be rushed, says Dienes. Children need plenty of time to experience the objects around them before one can or should shape the way they think about them.

Following this period of free play comes a period during which children's experiences can begin to be structured systematically using the concrete materials. The teacher, or perhaps an individualized mathematics program, can lead children through a series of learning experiences that point to the properties of interest and sharpen up the concept that is being learned. This is where the special characteristics of the manipulative math materials have their greatest impact upon learning. Mathematical games (see Holt & Dienes, 1973) are also helpful at this point, because the "rules" of games represent realistic restrictions on possible mathematical operations and thus help to guide and shape children's mathematical understanding.

What the learner does during this period of structured play, according to Dienes, is to begin to *abstract* a concept. To abstract a concept, a child needs to (1) "gather together that which is common to a large variety of experiences;" and (2) "reject that which is irrelevant to these experiences [Dienes & Golding, 1971, p. 95]." The process of abstraction goes on all the time in learning, but for the purpose of teaching higher-level mathematical concepts to young children it may need to be promoted through the artifice of concrete math materials such as we have described. To take a very simple example, in a game with attribute blocks children are led to abstract a principle—the set of pieces having in common the property of redness—when the teacher places before them, serially, a large, thick, red circle; a small, thin, red square; a large, thin, red triangle; etc.

In order for mathematical concepts to be properly abstracted from a series of learning episodes, says Dienes, the concepts must be presented in *multiple embodiments,* that is, children should work with several different kinds of materials, all of which embody the concept of interest. Multiple embodiments are viewed as facilitating the sorting and classifying process that constitutes the abstraction of a concept. Seeing a principle operating similarly even when different materials are used seems to help children discover what is and is not relevant to the concept. For example, suppose Jenny works extensively with attribute blocks to learn elementary set theory. She may develop the mistaken notion that the properties of sets are somehow related to the blocks' plastic quality or to their specific characteristics of size, shape, and color. Her error may be forestalled, or corrected, by having her also experience set theory using number-based materials such as Unifix cubes or number analogs like Cuisenaire rods. Underlying the notion of presenting concepts in multiple embodiments is an assumption that children are familiar with the various math materials from previous unstructured play. Without this prior familiarization, they may be required to learn the materials *and* a new mathematical principle at the same time, which might be counterproductive.

According to Dienes, the various embodiments should look as different from each other as possible—the principle of *perceptual variability*—so the children can "see" the structure from several different perspectives and build up a rich store of mental images surrounding each concept. Varying the embodiments perceptually is supposed to let the concept develop independent of the specific forms of materials. In Fig. 5.5 we saw two varieties of base 3 MAB—triangles and cubes—that differed visually but embodied the same principles.

Multiple embodiments should also permit manipulation of the full range of mathematical variables associated with a concept—Dienes' principle of *mathematical variability*. Mathematical variations are supposed to make clear the extent to which a concept can be generalized to other contexts. For example, the learning of positional notation need not be confined to the context of base 10 materials. Children familiar with other base systems can see that similar patterns of notation appear when using base 4 or base 6. Further, the materials should allow children to discover that they can perform number operations the same way regardless of the particular numbers being used. Whenever possible, Dienes believes concepts should be dealt with in their geometric, physical, and even social forms as well as their arithmetic and algebraic forms. To take a social example of a mathematical principle, Dienes has used a game in which the seating of four children around a table is rotated as an embodiment of the properties of the set of rotations of a tetrahedron (Dienes & Golding, 1971).

Continuing with the learning cycle, once children have been led through increasingly controlled manipulations or games with multiple embodiments of concepts, it is time to help them find ways to talk about their findings. According to Dienes, the next step is to encourage children to abstract their learning further away from the concrete materials by drawing simple pictures, graphs, or maps and eventually by attaching mathematical symbols to the concepts. The use of symbols should be informal at first, geared to helping children keep track of patterns and relationships they have noted. Children can even use symbols of their own choosing. This is thought to be a way of letting children participate in the exciting process of discovery and formalization that mathematicians experience. It also keeps the learning experience from becoming just another memorizing exercise.

The importance of symbolization is that it elevates mathematical activity to a new plane. Experiences up to this point are assumed to have been registered either as physical manipulations or as mental pictures of these manipulations and their outcomes (cf. Bruner's enactive and iconic representations, p. 112). Indeed, one of the stated functions of multiple embodiments is to provide a rich store of mental images. The transition to symbolic representation should ensure that these images will eventually be called up by the mathematical symbols that are attached to them (Dienes, 1963, p. 163). As symbols are applied, mathematical experiences are freed from their concrete referents and become the tools for new kinds of mental manipulations.

From this point on in the learning cycle, the learners' role is to systematize their learning, with the teacher's guidance where necessary, into a structured body of rules, albeit on a very small scale. Children now play with symbols and rules rather than with concrete embodiments, and they discover which manipulations and groupings of rules are possible. A new phase of free play is entered, now using symbols as the objects of manipulation and leading to higher-order structures of mathematical thought. However, Dienes (1963) cautions that symbols may tend to be manipulated without reference to the "reality" they symbolize unless "feedback is applied from time to time [p. 162]." Thus, children must be able to "revisit" the concrete manipulations phase, or at least images of it, at any time in order for the symbolism to remain vitally connected to their concrete experiences.

QUESTIONS RAISED BY THE STRUCTURE-ORIENTED APPROACHES

How does one evaluate the instructional principles of structure-oriented educators and researchers such as Bruner and Dienes? We have presented portions of their work as an example of the kinds of efforts arising from the ambiance of the early 1960s, when an interest in the new curriculum goals and an interest in applying psychological findings to instruction seemed to converge. These efforts are difficult to evaluate rigorously, however, because much of the research was informal and practice-oriented. Strictly controlled experiments were not possible due to the nature of these explorations, nor were detailed theories of mathematical thought processes developed. The researchers involved were primarily interested in learning whether children could be taught the more advanced topics of mathematics and only secondarily in explaining or defining the psychological processes of learners. However, their experimental programs, and the instructional materials and principles that emerge from them, do raise some important questions that any attempt to design mathematics instruction must consider and which therefore concern us throughout the remaining chapters.

The difficulty of evaluating these principles and instructional sequences can be attributed in part to a lack of rigorous evaluation criteria. Informal observations of the behavior of individuals and groups of children on mathematical tasks were the principal indicators of whether or not they had successfully learned concepts. Children's verbalizations were another clue, but verbalizations may not reflect the full extent of children's understanding. As Piaget has pointed out—and proponents of concrete math materials seem to subscribe to this view—children may "know" more than they are able to talk about. It seems clear that informal observations of behavior and verbalizations are not enough to permit thorough evaluation of a conceptual approach (or any approach) to instruction. We need to know what exactly constitutes "understanding" of mathematical structures. This question may be put in two ways: (1) Can we define psychologically the structures that we wish to teach? (2) Can we determine the nature of the learner's knowledge both before and after a particular teaching sequence so as to determine whether those structures have been learned? Both are very difficult questions to which psychologists have been devoting much time and effort. As cognitive psychology has developed, and particularly as the study of language understanding has progressed, new tools for mapping out the content of people's knowledge are slowly becoming available and perhaps may now be employed to evaluate structure-oriented teaching efforts. This line of questioning is pursued in Chapter 8, where we present recent attempts to analyze "understanding" and to define its relationship to the structures of mathematics.

Another evaluation issue has to do with determining whether mathematical structures are actually being taught in the structure-oriented approaches. First, mathematicians differ in their views of the subject matter and what is important for children to learn. Dienes (1967), for example, developed an "operational" definition of fractions in one of his lesson sequences; but geometric approaches to fractions—pieces of pie, etc.—have been more widely used as conceptual presentations, and one might argue that they are less confusing, more "meaningful," to children. Greeno (1976) has suggested that operational, geometric, and algebraic presentations of fractions may result in qualitatively different learning results and may be more or less transferable to other mathematical problems and tasks. Second, when one teaches a part of a larger mathematical structure, even in a mathematically "honest" way, is the partial structure not qualitatively different from the way it is later to be understood as a component of the whole mathematical edifice? In other words, can we be sure that such an approach is not merely creating double work for the child? Will the child learn a concept one way the first time and have to relearn it in another context later? Third, when we speak of learning a mathematical structure, do we mean mathematics as a mathematician might define it? Is it the formal mathematical structure that is important for a child to grasp, or is it more important for the child to build up a good intuitive psychological structure, that is, an organized set of associations, propositions, or relations that allows the child to use and acquire mathematical knowledge efficiently and flexibly?

Perhaps the most important evaluation question is whether teaching the structures of mathematics, using concrete materials in multiple embodiments, brings about better learning and deeper understanding of mathematics than traditional computational approaches. This is difficult to answer in the absence of definitive studies, because the evidence we have is largely indirect, based on experience in the schools. We can see in classroom teaching the results of the curriculum reform movement of the 1960s. Today many of the leading math textbooks emphasize concepts, and several devote more space to concept development than to computational skills (Scandura, 1971). Drill, which was once so widespread in mathematics education, is now sometimes relegated to a supplementary role, appearing in practice sections at the backs of textbooks. Nevertheless, we have not proved that teaching mathematical structures has improved children's ability to learn more complex topics later or to solve problems and think logically. This is because we simply do not know to what extent structure-oriented approaches have been implemented in classrooms and how student performance has been affected by this orientation to teaching. We know that children are now exposed to a wider range of mathematical content than 30 years ago. But we do not know to what extent classroom teaching has included opportunities for manipulation of concrete objects, whether teaching sequences have included multiple embodiments of concepts, and whether children have been led to an intuitive understanding of concepts before being required to verbalize their knowledge in mathematically correct terms. We cannot judge the success of this approach, and especially its success in comparison to computational approaches, until we can be sure it has been implemented in classrooms.

It seems certain that any attempt to teach mathematical structures cannot neglect the need for practice in computational algorithms and number facts. It also seems certain that instructional planning should include opportunities for learning both concepts and computation skills. It is entirely possible that structure-oriented approaches and the spiral curriculum notion have emphasized concepts at the expense of computational practice. If we admit the possibility that computational practice, accompanied or followed by presentations of structural principles, may even serve an essential role in the development of understanding, then perhaps the careful development of concepts before introducing procedures and algorithms may actually be counterproductive. These are issues that demand clarification through research. They may be particularly important in view of recent public concern over children's poor school achievement, and the resulting movement to return to more traditional forms of mathematics instruction that emphasize computational drill and practice over conceptual approaches. This movement may be unfortunate, because we are only now beginning to have the psychological tools for evaluating structure-oriented instruction, and so the definitive research that would permit comparisons between modes of teaching is just becoming possible.

In view of the many questions that remain unanswered concerning structure-oriented instruction and the work of Bruner, Dienes, and others like them, one is tempted to disregard this approach on the grounds that it has not demonstrated its own validity. Nevertheless, the approach has a certain appeal that demands a closer look. There are two reasons: First, this work shows a respect for children's intelligence and their capacity for inquiry and invention. Second, its objective is to convey the subject of mathematics to young children simply and elegantly. We may quarrel, perhaps, with people's ideas of which structures are important to teach and with how far one needs to go with the complex topics; but the basic aim of conveying the subject matter in a mathematically meaningful way and taking into account the cognitive capacities of children is a noteworthy goal and deserves continuing research by both psychologists and mathematics educators.

SUMMARY

During the decade of the 1960s a period of curriculum reevaluation and reform took place that had as its foundation a concern with teaching children the fundamental structures of mathematics and teaching them in a way that would take into account children's intellectual capabilities and motivational needs. In this chapter we have sampled some of the math materials and teaching strategies that were developed during this period, and we have presented some of the psychological underpinnings of structure-based mathematics instruction. Structure-oriented psychologists, educators, and mathematicians believe young children are capable of grasping more complex mathematical concepts than had previously been thought possible. They suggest that the structures of mathematics may be taught in an intellectually honest way at an early age by presenting them in concrete form, especially in the form of math materials that physically embody those structures. An understanding of the mathematical structures underlying the procedures and concepts taught in the classroom is seen as fundamental to meaningful learning.

From the field of cognitive psychology come suggestions for how instruction can be made responsive to learners' cognitive processes. Bruner, a psychologist associated with the curriculum reform movement, has a cognitive theory of conceptual development that implies a certain sequence of instruction. He asserts that mathematical structures can be built up in the minds of learners by providing experiences that allow them to develop enactive, iconic, and symbolic representations of concepts, in that order. These mental representations are hypothesized to be the forms or modes in which learning experiences, and ultimately concepts, are remembered.

Dienes, a mathematics educator, focuses on the use of concrete math materials in a similar sequence of learning experiences, a learning cycle. He suggests that structural concepts are discovered and refined as children engage in guided manipulations of materials that physically embody the concepts in several forms. Instruction and practice may be organized to highlight the distinctions between relevant and nonrelevant aspects of concepts and to expose children to the full range of perceptual and mathematical variations of those concepts.

The structure-oriented materials and instructional principles raise questions about the nature of mathematical understanding and about the sorts of structures they really teach. To evaluate this approach one would have to be able to define psychologically the mathematical structures one wished to teach and also be able to assess the degree of understanding a learner had before and after instruction. The structure-oriented methods and materials have not been adequately validated by research, and we know little from school practice about the effects of the curriculum reforms upon the quality of children's mathematical learning. The psychological tools for designing the needed research are only now becoming available. Nevertheless, the structure-oriented approach holds out a worthy goal for mathematics instruction—the design of teaching that presents the underlying structures of mathematics elegantly and simply while at the same time taking into account the cognitive capabilities of learners.

REFERENCES

Bartlett, F. C. *Remembering*. London: Cambridge University Press, 1932.
Bruner, J. S. *The process of education*. Cambridge, Mass.: Harvard University Press, 1960.
Bruner, J. S. The course of cognitive growth. *American Psychologist*, 1964, *19*, 1–15. (a)
Bruner, J. S. Some theorems on instruction illustrated with reference to mathematics. *The Sixty-third Yearbook of the National Society for the Study of Education* (Pt. 1), 1964, *63*, 306–335. (b)
Bruner, J. S. *Toward a theory of instruction*. Cambridge, Mass.: Harvard University Press, 1966.
Bruner, J. S., Goodnow, J. J., & Austin, G. A. *A study of thinking*. New York: Wiley, 1956.
Bruner, J. S., Olver, R. R., Greenfield, P. M., et al. *Studies in cognitive growth*. New York: Wiley, 1966.
Dienes, Z. P. *Building up mathematics*. New York: Hutchinson Educational Ltd., 1960.
Dienes, Z. P. *An experimental study of mathematics learning*. New York: Hutchinson & Co., Ltd., 1963.
Dienes, Z. P. *Mathematics in the primary school*. London: Macmillan, 1966.
Dienes, Z. P. *Fractions: An operational approach*. New York: Herder & Herder, 1967.
Dienes, Z. P., & Golding, E. W. *Approach to modern mathematics*. New York: Herder & Herder, 1971.
Dilley, C. A., Rucker, W. E., & Jackson, A. E. *Heath elementary mathematics*. Lexington, Mass.: D. C. Heath & Co., 1975.
Gattegno, C. *For the teaching of elementary mathematics*. Mt. Vernon, N.Y.: Cuisenaire Company of America, Inc., 1963.
Goals for mathematical education of elementary school teachers: A report of the Cambridge Conference on Teacher Training. Newton, Mass.: Education Development Center, 1967.
Goals for school mathematics: The report of the Cambridge Conference on School Mathematics. Boston: Educational Services (by Houghton Mifflin), 1963.
Greeno, J. G. Cognitive objectives of instruction: Theory of knowledge for solving problems and answering questions. David Klahr (Ed.), *Cognition and instruction*. Hillsdale, N.J.: Lawrence Erlbaum Associates, 1976.
Holt, M., & Dienes, Z. P. *Let's play math*. New York: Walker, 1973.
Koffka, K. [*The growth of the mind*] (R. M. Ogden, trans.). London: Kegan Paul, Trench, Trubner, 1924.
Köhler, W. *The mentality of apes*. New York: Harcourt, Brace & World, 1925.
Montessori, M. *Advanced Montessori method*. Cambridge, Mass.: Robert Bentley, 1964. (Reprint of original publication, 1917)
Payne, J. N., & Rathmell, E. C. Number and numeration. In J. N. Payne (Ed.), *Mathematics learning in early childhood*. Reston, Va.: National Council of Teachers of Mathematics, 1975.
Piaget, J. *The child's conception of number*. New York: Norton, 1952. (Original French edition, 1941.)
Rasmussen, L., Hightower, R., & Rasmussen, P. *Mathematics for the primary teacher* (Mathematics Laboratory Materials, Primary Edition). Chicago: Learning Materials, Inc., 1964.
Scandura, J. M. *Mathematics: Concrete behavioral foundations*. New York: Harper & Row, 1971.
Trafton, P. The curriculum. In J. N. Payne (Ed.), *Mathematics learning in early childhood*. Reston, Va.: National Council of Teachers of Mathematics, 1975.

6 Structure and Insight in Problem Solving

Long before the Cambridge and Woods Hole Conferences articulated the need for teaching mathematical structures in meaningful ways, another group of psychologists in Europe was developing a theory that also pointed to the importance of understanding structure for problem solving and thinking in general. These were the gestalt psychologists. Although during the first half of this century a few psychologists such as Brownell were concerned with meaningful conceptual learning, most American psychologists of the time were still largely involved in accounting for the formation of simple associations. Thorndike acknowledged the necessity for more complex organizations of bonds to account for problem solving and understanding, but his experiments remained focused upon the basic connections of which, he thought, all knowledge consisted. Imported from Europe in the 1920s, gestalt psychology set forth a fundamentally different approach to learning, citing experimental data that the associationist theories could not readily explain. Gestalt theory influenced portions of American psychology for years afterward and today is receiving renewed attention in analyses of problem solving.

In this chapter we present a brief introduction to gestalt psychology. We consider the work of Max Wertheimer, a gestalt psychologist who was particularly concerned with mathematics learning and teaching and who claimed to demonstrate in dramatic fashion the different results that could be expected from rote learning and learning with meaning. We then look at gestalt explanations of problem solving, focusing particularly on the phenomenon of insight and the processes by which it might be achieved. A gestalt view of learning is presented and related to more recent studies on "discovery" learning. Finally, we explore the implications of a gestalt approach for the teaching of mathematics and problem solving.

GESTALT PRINCIPLES AND SOME MATHEMATICAL EXAMPLES

Gestalt theory grew out of a European tradition of psychology that accepted phenomenological reports (individuals' descriptions of their mental experience during some task) as basic data and sources of hypotheses. Gestalt psychologists—notably Köhler, Koffka, and Wertheimer—were distinguished by their insistence that the human mind *interprets* all incoming sensations and experiences according to certain organizing principles, so that, rather than merely taking in the information, some sort of understanding is achieved. They sought the instances in everyday life that would support the existence of the hypothesized organizing principles. Central to their style was the arranging of demonstrations in which they believed the operation of the theory was dramatically and indisputably revealed.

The earliest gestalt work was concerned with the organization of human perceptual processes. According to the theory, human perception could not be accounted for merely as a summation of all the stimuli that impinge upon the senses. This presented a clear contrast with the extreme reductionism of American psychology of the time. Gestalt theory held that the perceiver brought something unique to the experience of perceiving, something that made the experience more than just the sum of its constituent stimuli. There was a natural tendency for the perceiver to see structure in his or her perceptions. There was a pressure to seek whole forms or "gestalts" in the environment, and this was affected by specific rules of perception.

Some of the gestalt notions are exemplified in Fig. 6.1. Notice the configuration of dots labeled (a) (see Luchins & Luchins, 1970). A typical reaction to this pattern is that it forms a diamond shape. But if one adds two more dots as we have done in (b), the pattern forms a triangle, with the topmost dot functioning as its vertex. Now, if one erases the three bottom dots and adds two above as in (c), the result is a rectangle, in which our pinpointed dot marks the center. In (d), the dot is perceived as merely peripheral to another visual pattern, the hexagon. This simple example demonstrates how the identity and function of a perceptual component, in this case a dot, changes with its surrounding context. One actually sees the dot differently depending on its relation to other dots around it. Thus, context determines the way things are perceived.

Furthermore, we do not think of what we see in Fig. 6.1 as merely a collection of dots. According to gestalt theory, our perception of the dots is dominated by our tendency to see them as groupings that we recognize as shapes—diamond,

130   6. STRUCTURE AND INSIGHT IN PROBLEM SOLVING

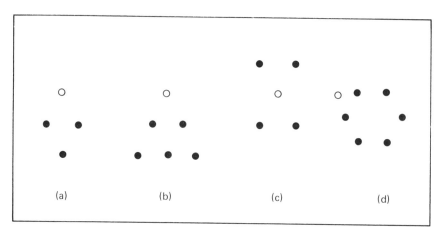

FIG. 6.1   Dominance of perceptual organization. A dot (the empty circle) is experienced differently according to the sounding context.

triangle, square, and hexagon. Again, the perceiver does not merely register individual dots; he or she brings to the experience an organizing principle that makes the whole perception add up to more than the sum of its parts.

Gestalt psychologists cited numerous demonstrations of this phenomenon in everyday experience. Music was a good example, because melody was composed of many individual notes, but when those individual notes were transposed to a new key, they retained their character as a melody. It was their interrelations, or the pattern of the notes, that the listener perceived and not the individual tones. The structure contributed by the subjective experience of perception was further demonstrated in the phenomenon of apparent motion (Wertheimer, 1923, translated and reprinted in 1938). A movie film was made up of thousands of still frames, but when these were presented in fast succession, one perceived the sum of the individual frames as a moving picture. As a final example, notice the shapes in Fig. 6.2. These are broken or incomplete shapes, but, rather than seeing them as a collection of connected lines and curves, we perceive them as whole forms with gaps and additional pieces. Perception tends to seek "closure" in such figures. The tension created by the visual incongruity is resolved into the perception of a unified whole.

Although concerned initially with perceptual phenomena such as we have demonstrated, the attention of gestalt psychologists eventually focused on a more general problem, the nature of thinking and problem solving. The gestaltists came to believe that thinking and perceptual processes were governed by the same basic principles. They suggested that the way patterns in visual and auditory arrays were registered by the perceiver might be very much like the way thoughts were organized by the thinker. In other words, the psychological field (the "inner space" in which cognition occurred) might be subject to the same tendency to see structure. If so, then thinking too would be affected by context, and incongruities among ideas would seek "equilibrium" in pure structural forms (an analog of closure in perception).

Insight and Problem Structure. One focus of gestalt studies in problem solving was the phenomenon of insight. The role of insight in problem solving was introduced to the gestaltists largely through the efforts of Wolfgang Köhler (1925), who had worked closely with Wertheimer on the early experiments in perception. Köhler had occasion to observe closely the behavior of a captive colony of chimpanzees over a number of years. He especially noted their efforts to solve everyday problems, such as trying to obtain food that was just out of reach. Whereas learning theorists like Thorndike tried to explain animal problem solving in terms of goal-oriented trials and errors, Köhler's observations convinced him that more global organizing processes were at work. Behavior was not always directed strictly toward the goal object; solutions sometimes involved temporarily turning away from the blocked goal and searching for a detour that would ultimately lead to the same end. For example, finding that a banana was high up on a shelf, an ape would turn away from the food and move to the other side of the cage to procure a box to stand on, demonstrating a human-like "reasoned" response to a problematic situation.

Such instances seemed to follow moments of apparent insight into some important aspect of the problem. Recognition of the nature of the problem and of the solution often happened suddenly and simultaneously after a long period of

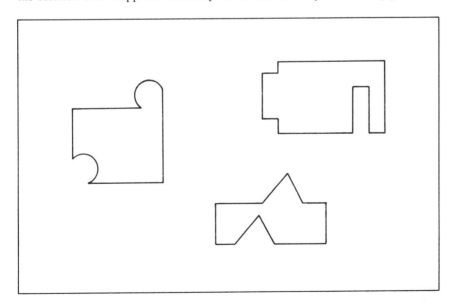

FIG. 6.2 Shapes seen as unified figures with "extras" and "gaps."

132 6. STRUCTURE AND INSIGHT IN PROBLEM SOLVING

seemingly random activity. To cite another example, an ape was given two sticks to play with, one of which could be fit into the other to make a single longer stick. A banana lay on the floor just out of reach outside the cage. After a period of play with the sticks, the ape suddenly seemed to understand how they might be used to solve the problem. Then, with deliberation, the ape fitted the sticks together and used the elongated stick to bring the banana within grasping range. Given such behavior, Köhler reasoned, thinking and problem solving could not be the mere sum of constituent stimulus–response associations but must involve perceiving problems as functional wholes. This understanding or insight, interpreted as recognition of the problem structure, seemed to result from a reorganization of problem elements so that they were seen in a new context. The relation of the box or sticks to the configuration of elements that constituted the "banana problem" suddenly became clear to the ape, and solution behavior followed almost immediately.

Extrapolating to human behavior, Köhler viewed problem situations as creating tensions in a psychological field, much as the blocked goal initiated solution activity in apes. Where a solution was not immediately apparent where there was inherent conflict in the problem situation, the dynamic mental forces would seek equilibrium in the form of some reorganization. This reorganization would resolve the tension (i.e., reveal the true structure of the problem and thus the path to solution). According to Köhler, the resolution of this inner conflict could only come about when the components of the problem were perceived in their proper function with regard to the whole. Insight occurred at the point of this reorganization. The structure of the problem thus revealed would define the functions and interrelationships of problem elements and consequently determine which skills could be applied toward solution. Until the basic structure of the problem was apprehended during this moment of insight, the problem situation was not meaningful to the would-be problem solver and the problem was, therefore, not solvable.

Wertheimer's Parallelogram Problem

What the gestalt notion of structure in human cognition might mean with respect to education was perhaps best exemplified in the work of Wertheimer. Although Wertheimer began by studying perceptual phenomena in the laboratory, he became increasingly interested in human thinking. Eventually he moved out of the laboratory and into the classroom to study thought processes in children. Wertheimer was particularly interested in demonstrating what he called "productive thinking" or thinking based on an appreciation of structure. In a favorite demonstration, the parallelogram problem, Wertheimer claimed to reveal the operation of productive thinking and suggested what such thinking entailed.

Wertheimer (1945/1959) tells of going into a schoolroom where young children were being taught to find the area of a parallelogram. Their teacher had shown them how to construct a line (drop a perpendicular) from the upper left-hand corner such that it formed a 90-degree angle with the base. They were then to measure the new line and multiply it by the length of the base to find the answer (see Fig. 6.3, top). Using this standard algorithm, the children were successfully computing the areas of many practice figures when Wertheimer stepped to the front of the room and posed a disturbing problem. He showed them the parallelogram depicted at the bottom of Fig. 6.3 (an "upended" version of the one at the top of the figure), and asked them to find its area.

The children's reactions were mixed. Some declared "no fair," saying they had not been taught how to do that kind of problem; others said their altitude-times-base algorithm simply would not work on that kind of figure; still others simply gave up and refused to consider the problem at all. The difficulty was that when the children dropped a perpendicular from the top left-hand corner, as they

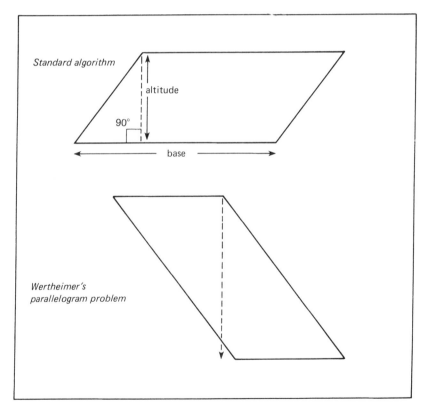

FIG. 6.3 Finding the area of a parallelogram using the standard algorithm. Children were confused when applying it to Wertheimer's problem figure. (Adapted from Wertheimer, 1945/1959.)

134   6. STRUCTURE AND INSIGHT IN PROBLEM SOLVING

had been taught, the line ended up somewhere to the left of the base, so that the standard formula did not seem to apply (see Fig. 6.3., bottom).

Let us analyze the parallelogram problem from the point of view of mathematical structure. What do the children actually do when they drop a perpendicular, measure it, and multiply by the base? They apply a general algorithm for calculating the area of a four-sided figure; that is, multiplying base times altitude is a shortcut method for dividing the figure into a number of uniform squares and counting them to determine the number of "square" inches. Understanding the nature of this algorithm, one should have no trouble figuring out how to apply it to squares or other rectangles. But the algorithm's relationship to the parallelogram is not so obvious, at least not from the standpoint of its perceptual impact. The human tendency to perceive things as organized wholes—the parallelogram as a unified figure—may in this case obscure the true structure of the problem. Look again at the drawings in Fig. 6.2. Each drawing tends to be seen as a rectangle with a "gap" and an "extra" piece that could fill the gap. Now, suppose one were to view the parallelogram in terms of the gap-extra relationship. It could be perceived as another shape, a rectangle, with a piece sticking out that fits perfectly into a gap on the other end. Next notice that if a parallelogram is transformed by dividing it along a perpendicular and refitting the pieces such that they form a rectangle, the altitude-times-base formula applies quite directly. This explains why one drops a perpendicular in executing the area algorithm.

The point Wertheimer wished to make with his demonstration was this: Children who learned the algorithm for finding area without understanding the structural principles upon which it was based were limited to following blindly the rules set forth by their teacher. They had obviously been taught the algorithm in a rote fashion; they had not learned with meaning. This was of great concern to Wertheimer, because he felt that the schools of his time were instilling the habit of applying algorithms in a senseless fashion and thus stamping out children's natural human tendency to see things as structured wholes. Blind algorithmic solutions he called "ugly" and "foolish." Solutions derived from a true understanding of the problem structure, on the other hand, he variously called "elegant, beautiful, true, and clean" and saw as examples of productive thinking.

In Wertheimer's "problem" parallelogram, the difficulty was that "rule-bound" children made their lines perpendicular to the bottom of the page rather than to the base of the figure. They would not have made such a mistake, said Wertheimer, if they had understood the functional equivalence between the parallelogram and the rectangle. The equivalence was the true underlying structure of the problem. Conceiving the problem in that light was a way of organizing "sensibly" with respect to both the goal of the task (finding area) and the basic perceptual and mathematical features of the parallelogram. It took into account both the context of the problem and the relation of the parts to the whole. One could, of course, argue that the children had learned a wrong algorithm or had at least been taught an incorrect definition for the altitude of a figure. Had the children been given correct rules to follow, their strict adherence to an algorithm might have yielded performances indistinguishable from those of children who understood the underlying mathematical structure.

Further, in the foregoing example it is not clear that the gestalt tendency to perceive things as organized wholes necessarily facilitated problem solving. In fact, we saw that the perceptual wholeness of the parallelogram might interfere with noticing the gap-extra relationship that transformed the figure into a rectangle to which the standard formula for area applied. In this and other gestalt demonstrations, the perceptual structures and the true structures of the problems often seem to vie for the mind's attention. This presents a persistent difficulty in interpreting gestalt theory with respect to problem solving. Nevertheless, Wertheimer finds ample evidence for the distinction between productive and nonproductive thinking, and he cites these differences as the differential results to be expected from rote as opposed to meaningful learning.

The Carpenter's Apprentice and the Sum of a Series

An illustration of the two supposed approaches to a problem—one based on an understanding of structure, the other a blindly algorithmic solution—is found in Wertheimer's demonstration of the derivation of Gauss's formula for the sum of a series. According to mathematical lore (Bell, 1977), young Gauss, a boy of 10, discovered the principle behind his famous formula one day when a sadistic teacher set his class the problem of finding the sum of a series of large numbers and sat down to read the newspaper while they worked. Within minutes, the story goes, young Gauss had the answer—and a very astonished teacher! Gauss hadn't wanted to waste a lot of time adding each successive number, he said, so he had found a shortcut.

On first thought the task of summing a series of numbers, say from 1 to 100, indeed appears to require a long process of computation: $1 + 2 + 3 + \ldots + 100$. But, as Gauss found out, there is a simpler and more elegant way to find the sum. Although the exact thinking behind his discovery is not known, Wertheimer uses a variant story as a point of departure for his own inductive exposition of the properties of the series. We follow the steps of Wertheimer's reasoning in order to give a flavor of the structure-based insights he thought Gauss might have had in examining this task (Wertheimer, 1945/1959, pp. 108–116):

> A staircase is being built along the wall in the hall of a new house. It has 19 steps. The side away from the wall is to be faced with square panels of the size of the ends of the steps. The carpenter tells his apprentice to fetch them from the shop. The apprentice asks, "How many panels shall I bring?" "Find out for yourself," rejoins the carpenter. The apprentice starts counting: $1 + 2 = 3; + 3 = 6; + 4 = 10; + 5 = \ldots$.

136   6. STRUCTURE AND INSIGHT IN PROBLEM SOLVING

The carpenter laughs. "Why don't you think? Must you count them out, one by one?"

Dear reader, if you were the apprentice, what would you do?

If you do not succeed in finding a better way, I will ask: "What if the staircase were not along the wall and required the square wooden panels on both sides? Would it help if I suggested thinking of the patterns of the two sides cut out of paper?" [Wertheimer, 1945/1959, p. 108].

Wertheimer is suggesting that the reader visualize the shape of the staircase panels as half of a larger figure. Fitting the side of the staircase together with another piece exactly the same shape and size yields a rectangle. Thus conceived, the problem can be represented as in Fig. 6.4. Viewing the staircase as half of a rectangle and filling in the gap temporarily with an imagined complementary staircase shape, we have transformed the problem. Instead of looking like a

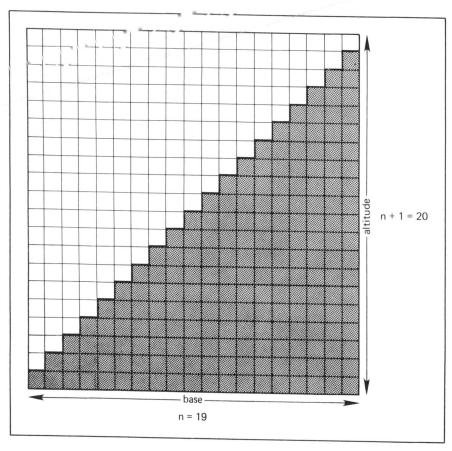

FIG. 6.4 Visualizing the staircase problem as two halves of a single figure. (Adapted from Wertheimer, 1945/1959.)

Gestalt Principles and Some Mathematical Examples   137

tedious computation of successive additions, the problem now appears to yield to a simple two-step multiplication and division process—finding the number of the squares in the whole rectangle and then dividing by 2 to find the number of outside panels in the staircase.

How does this intuition relate to the problem of summing the numbers from 1 to 100? Instead of computing 1 + 2 + 3, etc., as the carpenter's apprentice might have done, Wertheimer suggests we bring the problem into the visual domain to try to understand its structural properties.

```
1  2  3  4  . . .  97  98  99  100
```

Note that the numbers can be paired in such a way that each pair sums to the same number. Note also the number of pairs this procedure would yield and the relation of the number of pairs to the total number of elements in the series. Fifty pairs, each summing to 101, gives us the answer, 5050.

Another diagrammatic approach Wertheimer points out also involves pairing but is structurally different from the approach above. It is directly analogous to the strategy suggested for solving the staircase problem. We imagine a complementary series of identical numbers in order to simplify the computation.

```
  1    2    3    4   . . .    97   98   99   100
100   99   98   97   . . .     4    3    2     1
```

Now we have two complete series of 1 to 100. They are arranged in pairs, each of which sums to 101. One-hundred pairs of 101 equal 10,100; then since we began by doubling the series, we divide the product in half to find the desired sum, 5050.

Of course we need not write out all the pairs in the manner shown above. The point of these demonstrations is simply to capture the problem structure intuitively. Once we understand the principle of pairing, and the invariability of the pair sums, we can even invent shortcut formulas to describe the operations we have performed on the series. The results are two versions of Gauss's formula for the sum of an even-numbered series, either $\frac{n}{2}(n + 1)$ as in the first example or $n\frac{n+1}{2}$ as in the second. According to Wertheimer, the two versions of the formula are mathematically equivalent but structurally and psychologically different. As it turns out, the formula applies equally well to odd-numbered series, such as the series 1–19 in the staircase problem. When we pair the numbers 1–19 as in the first example we have one number left over. This number, 10, we may simply view as "half a pair," says Wertheimer ($\frac{n}{2}$)

What Wertheimer hoped to achieve by this demonstration was the intuition behind Gauss's general formula. The structural properties of the series were apparent when it was visually represented, and this should have led quite readily to recognition of shortcuts to solution. We do not know that this is how Gauss saw the series problem, only that the intuition behind his formula was more easily grasped when the problem was recast in visual or diagrammatic terms. Wertheimer's demonstrations have great intuitive appeal. Yet one questions whether all or even most mathematical problems can be set up in visualizable forms. Mathematical structure is not always paralleled by spatial structure. It may be no accident, therefore, that many of Wertheimer's pedagogical examples are drawn from geometry, which deals with spatial rather than algebraic concepts.

One might also ask, why shouldn't the carpenter's apprentice have merely counted $(1 + 2 + 3 + \ldots + 19)$ the number of panels he would need to cover the side of the staircase? And why not do the same with the series of 1 to 100? The answers would have been just as valid as those arrived at by more structural conceptualizations. But such solutions would have been "ugly" according to Wertheimer's scheme of things. We can see that they would not have been based on an understanding of the structure (i.e., the mathematical structure) of the problems. They are also clumsy and inelegant by any mathematician's definition. The solutions that are presumed to reveal productive thinking are elegant because they organize the components of problems along structural lines and deal with them in a parsimonious way. In these cases an understanding of mathematical relationships leads to simplification of the problems. It also saves both time and computational steps. This, after all, was Gauss's intention in the first place and remains a prime motive for many mathematical discoveries. An added benefit of the reduction in steps, of course, is the lessened chance for computational errors.

Note, however, that once the formula for the sum of a series is derived, it can be applied in an algorithmic fashion just as a simple addition algorithm might have been applied by the apprentice. The addition procedure itself has an underlying mathematical structure—the additive property of the real-number system. Perhaps we should not conclude from Wertheimer's examples that algorithms themselves are at fault in inelegant thinking; rather, the important point is that algorithms should be learned in the context of the structures underlying them. According to this interpretation, teaching should attempt first to build an intuitive understanding of the mathematical substrate for any computational procedure and only then introduce the algorithm as a "shortcut" for the more elaborate thinking upon which the algorithm is based. The suggestion is that when the reasons behind algorithms are clearly understood, then the thinker or problem solver is in a better position to choose the particular algorithm that is most appropriate to the problem at hand.

THE PROCESS OF PRODUCTIVE THINKING

The gestalt notion that the structure of the whole defines the functions and interrelations of its parts seems particularly relevant to the development of problem solving and generalized thinking skills. We have seen examples of so-called productive thinking in Wertheimer's demonstrations in connection with the parallelogram and carpenter's apprentice problems. But we are left wondering how such thinking proceeds. Is there a step-by-step process we can follow in the mind of the problem solver? If problem structure is as important as Wertheimer suggests it is, can we describe the process by which the problem solver apprehends structure, by which he or she achieves insights? And since our concern is with instruction, can an understanding of the process of productive thinking help us make children better problem solvers?

With respect to these points, the gestalt writings are not very specific. Consider Wertheimer's (1945/1959) comments on a well-conceived solution to the parallelogram problem:

> [There was] regrouping with regard to the whole, reorganization, fitting; factors of interrelatedness and of inner requirements [were] discovered, realized and followed up. The steps were taken, the operations were clearly done in view of the whole figure and of the whole situation. They arose by virtue of their part-function, not by blind recall or blind trial; their content, their direction, their application grew out of the requirements of the problem. Such a process is not just a sum of several steps, not an aggregate of several opperations, but the growth of one line of thinking out of the gaps in the situation, out of the structural troubles and the desire to remedy them, to straighten out what was bad, to get at the good inner relatedness. It is not a process that moves from pieces to an aggregate, from below to above, but from above to below, from the nature of the structural trouble to the concrete steps [pp. 49-50].

Sorting out just how classroom pupils may be encouraged and helped to engage in such thinking is not an easy matter. The idea of problem structure, although it inspires eloquence in Wertheimer's writing, is not always simple to define in practical terms. It is easier to demonstrate the operation of structural notions in the context of specific problems than it is to state in the abstract exactly what these structures are. And it is harder still to give a complete theoretical presentation concerning structure as a general phenomenon in mathematical problem solving. To a great extent, the gestalt psychologists have left their explanations at the level of concrete example. We can see that they are getting at something important, but we are not shown how to generalize the findings into principles for instruction that can be applied to varieties of specific mathematical content and problems.

Wertheimer's quotation, effusive and vague, is nevertheless interesting because it sets up a dichotomy that has been taken up again in recent studies of problem solving. The dichotomy is between processing that moves "from below to above, from pieces to an aggregate," and processing that moves "from above to below, from the nature of the structural trouble to the concrete steps." In mathematical problem solving, going from pieces to an aggregate implies letting various problem features suggest specific known procedures and strategies; going from the aggregate to the specifics implies examining various characterizations of the problem through such strategies as goal analysis and problem reformulation.

In current cognitive psychology, an important distinction is drawn between processing that is "bottom-up" and "top-down." Clarification of this distinction seems in order because the two types of processing suggest different approaches to instruction. If processing proceeds from the bottom up, that is, from specific characteristics of the materials, then emphasis in teaching should be placed on basic number facts and simple calculations before moving "up" to abstract mathematical concepts. In contrast, top-down processing suggests an approach to mathematics instruction that initially emphasizes the logic and structure of mathematics and only later insists on the details of computational algorithms and other such components of specific problems. In practice, it is often difficult to sort out bottom-up from top-down processes; in mathematical performances the processes appear to interact constantly. Nevertheless, the distinction is interesting pedagogically because it suggests alternative instructional foci.

Duncker on Problem Solving

Among gestalt psychologists, Karl Duncker (1945) most explicitly pursued the distinctions between bottom-up and top-down processing and attempted to illustrate them in the context of specific problems. In so doing, he seemed to advocate precisely the kind of process analyses of problem-solving behavior that many cognitive psychologists are developing today.

Whereas Wertheimer was concerned with the mathematical structures underlying specific problems, Duncker, a student of Wertheimer's, concentrated on general strategies of problem solving. He sought to explain problem solving in terms of the sequence of events that occurred between recognizing a problem and finding its solution. His object of study was the behavior of adult subjects on a variety of mathematical and practical problems (such as proving there is an infinite number of prime numbers, finding a way to irradiate a tumor without destroying healthy tissue, and building a door that opens from both sides). In problems such as these, where complex reasoning was required and a lengthy search was necessary to find a workable solution, problems appeared to undergo a series of reformulations in order to bring to mind solutions that had a functional relationship to the problem structure. Duncker (1945) hypothesized that the process was as follows:

> The final form of an individual solution is, in general, not reached by a single step from the original setting of the problem; on the contrary, the principle, the functional value of the solution, typically arises first, and the final form of the solution in question develops only as this principle becomes successively more and more concrete. In other words, the general or "essential" properties of a solution genetically precede the specific properties; the latter are developed out of the former [pp. 7–8].

He noted that a problem could be solved "from above" by reformulating the problem so that a particular class of solutions would be sought. This type of solution depended on what Duncker called an "analysis of the conflict," that is, figuring out what was wrong, what needed to be changed. "Analysis of goals," another form of solution from above, involved focusing on what the problem really demanded, so as to overcome the normal tendency to become fixed on particular lines of solution attempts. Solutions could also come "from below" through noticing features of the task and allowing those features to suggest possible solutions. This type of solution depended upon an "analysis of materials," that is, noting what was present and what could be used.

Duncker's hypotheses are illustrated in "talking aloud" protocols of adults trying to find an answer to the question, "Why are all six-place numbers of the form 267,267; 591,591; 112,112 divisible by 13?" The protocol of one individual was as follows:

1. Are the triplets themselves perhaps divisible by 13?
2. Is there perhaps some sort of rule here about the sum of the digits, as there is with divisibility by 9?
3. The thing must follow from a hidden common principle of structure—the first triplet is 10 times the second, 591,591 is 591 multiplied by 11, no: by 101. (E: So?) No: by 1001. Is 1001 divisible by 13? (Total duration 14 minutes) [Duncker, 1945, p. 31].

According to Duncker, the third solution strategy, looking for a hidden common principle, grows out of an analysis of the goal. What the subject has to discover is the general principle that all numbers of the form *abcabc* have the number 1001 as a factor, 1001 being divisible by 13. The reformulation of the problem by this subject creates a goal of finding a hidden common principle, setting up "conflict" that serves to direct and motivate activity to find that common principle. This can be characterized as an attempt at solution from above. But we see that the clue that leads to solution is actually a suggestion from below: The factor 1001 is discovered by noticing something about the nature of the six-digit numbers, namely, the consistent relationship of the first triplet to the second. The divisibility of 1001 by 13 is the only remaining discovery needed for problem solution, and this is quickly verified.

In an experiment devised to test the efficacy of suggestions from "above" and "below," Duncker gave different groups of subjects different hints while they worked on the "13" problem. The hints that markedly improved the likelihood of solution (producing a 50% solution rate) were: "The numbers are divisible by 1001" and "1001 is divisible by 13." Both are suggestions from below in the sense that they set the subject on the track of the number 1001, what it is divisible by, and what is divisible by it. A suggestion to analyze the goal ("Look for a more fundamental character from which the divisibility by 13 becomes evident") did not help and neither did general statements about the properties of division (e.g., "If a common divisor of numbers is divisible by 13, then they are all divisible by 13"). Although bringing attention to the number 1001 was what worked in this problem, Duncker found that the number need not have been mentioned explicitly, had the experiment been set up in such a way as to facilitate its discovery. When the problem was presented using successive numbers as examples—"Why are six-digit numbers of the form 276,276; 277,277; 278,278 always divisible by 13?"—most subjects subtracted successive numbers from each other (again, an attempt to find the hidden underlying principle), thus arriving at the number 1001.

Although Duncker expressed a preference for solutions from above based on analysis of goals and analysis of conflict—he called such solutions "organic" as opposed to "mechanical"—the details of his studies fail to establish the superiority of one strategy over the other. It seems more useful to acknowledge the effects of both goal analysis and analysis of materials and to appreciate their complex interactions. In Chapter 8, we examine these same processes from the perspective of modern cognitive psychology. We describe experiments specifically designed to tease out the differential effects of task materials and general solution strategies, and we relate these effects to models of human information processing.

A Gestalt View of Learning

The principles set forth by Wertheimer and Duncker, among other gestaltists, shed light on some complex aspects of human thinking—the organization of perception and problem solving. But although both were interested in fostering elegant thinking, and although their demonstrations appeared relevant to teaching, their work did not extend very far into the realm of instruction. Katona (1940/1967) cast the gestalt concern with structure and meaning into a more traditional experimental form by systematically comparing gestalt principles and traditional learning theory as they explain how people learn to perform tasks. Like Wertheimer, Katona labeled various types of learning either "senseless" or "meaningful." By senseless learning, he meant rote memorization. Meaningful learning, or learning by understanding, was based upon organizing a set of structurally related ideas or components.

To test the differential effects of these two types of learning, Katona designed a series of studies to be conducted in several stages. The general idea was to look for tasks that could be taught by two or more methods based on different theoretical orientations. Each method brought about learning in the short run as measured by pretests and posttests. First, experimental subjects were pretested to be sure they were not already competent at the task to be taught. Different groups of experimental subjects were then given either instruction based on rote memorization or instruction that stressed the principles underlying the tasks. In the instructional conditions that stressed meaning, some groups had to detect the principles themselves and some groups were given explicit descriptions of the principles.

At the end of the period of instruction, all subjects could usually do the task, because the various teaching procedures were carefully chosen and applied to ensure learning. The crucial comparison of learning types came at a subsequent stage. At this comparison stage, Katona tested subjects to see how well they could to the task a month later (retention). He also examined how learning one task contributed to executing other tasks that shared some characteristics with the one taught (transfer).

In one experiment, for example, Katona (1940/1967) asked people to learn a lengthy series of numbers, such as 1 4 9 1 6 2 5 3 6 4 9 6 4. Three different groups were given three different sets of instructions as follows:

Group 1. Recite these numbers slowly three times, for example, "one hundred forty-nine or one hundred sixty-two, ..."

Group 2. Read this slowly so you may know it completely and precisely: "The gross national product of the United States last year was $14,916,253,649.64."

Group 3. Try to learn the following series (i.e., no special instructions were given this group, just the printed list of numbers).

Group 3 subjects pondered the series briefly and then appeared to notice, or "discover," a certain pattern in the numbers: 1 4 9 16 25 36 49 64. The other two groups followed the specific instructions given.

Directly following the completion of this given task, all groups could recite the list of digits virtually without error. But during the comparison stage, differences emerged among the groups. Asked a week later whether they still remembered the series, Group 1 said the question was "unfair"; Group 2 remembered a partial answer, such as "the GNP was about $14 billion"; but Group 3 was able to remember the list perfectly and could in fact extend the series even further (e.g., 81 100 121, . . .). Thus, Katona reasoned, there was a qualitative as well as quantitative difference between the kinds of learning engaged in by the three groups.

Notice that some type of grouping was induced by each of the three forms of instruction. The kind used by Group 1 was clearly irrelevant and even interfered with noticing the pattern of the series. Group 2's partial answer was a sensible response to the memory question, given the context established by the original instructions. In fact, this may be interpreted as a structure-based response, albeit a response to structure of a different kind than that of a number series. Group 3, the "meaningful" learning group, was the most successful in the memory task. Having discovered the principle underlying the series, or in some cases having been told the principle, they grouped the numbers along structural lines, consistent with the underlying mathematical organization of the series. This group also demonstrated transfer in their ability to extend the series.

Katona's experiments were an attempt to prove that learning did not consist merely of memorizing a set of associations or a procedure. Learning could also mean reorganizing information so as to form a structure that had the power to

144   6. STRUCTURE AND INSIGHT IN PROBLEM SOLVING

explain other similarly structured problems. In Katona's view, this accounted for the transfer of knowledge to new situations. Finding the "problem structure"—the principle underlying the problem—not only made it easier to do similar problems but also enabled one to *reconstruct* the solution long after the initial exposure to the problem task. This was because the reorganization that accompanied meaningful learning provided lasting principles to guide reconstruction. In memorizing lists of digits, for example, persons who learned the principles for generating the series were able to reconstruct and extend them indefinitely. The influence of gestalt thinking is clear in Katona's explanations for such phenomena: Facts that are organized into a structured whole are retained as part of that whole, each being remembered because of its place within that structure.

Based on a series of similar experiments using card tricks and match-stick problems, Katona (1940/1967) reached several conclusions regarding the nature of meaningful learning:

> (a) Learning by memorizing is a different process from learning by understanding; (b) learning by understanding involves substantially the same process as does problem solving—the discovery of a principle; (c) both problem solving and meaningful learning consist primarily in changing, or organizing, the material. The role of organization is to establish, discover, or understand an intrinsic relationship [pp. 53–54].

Note that Katona, in his time, was reacting against a very doctrinaire S–R theory, as were Wertheimer and the other gestalt psychologists, hence his condemnation of learning by memorizing, which he apparently equated with rote learning. This argument has softened with time, and most psychologists today recognize that a sharp dichotomy between rote and meaningful learning is not warranted, that memorizing is not necessarily an *alternative* to understanding. Indeed, current work on memory makes it clear that memorizing is an active process, dependent on organizing principles very much like those Katona proposes. Research shows that a tendency to organize information as a means to more efficient memory increases with age, and this increase correlates with better performance on memory tasks (Kreutzer, Leonard, & Flavell, 1975). In this connection, it is interesting to note that people in Group 3 of Katona's number series task actually had less to remember than the other groups, because they only had to remember the *principle* for generating the series. Thus, organizing one's learning may lead to more efficient remembering partly because it decreases the number of separate pieces of information that must be retained. And organizing along structural lines, dictated in part by the structure of the subject matter, may be what leads to better retention in memory.

A Comment on Discovery Learning

Katona's conclusions twice mention the term *discovery,* indicating a shared concern with many later cognitive psychologists and educators. It is therefore of interest to note the parallels between Katona's conclusions and those arising from the extensive research on "discovery learning." Discovery has often been proposed as the best way to teach new concepts in mathematics and other subject areas. The strategy is to make available to children all the relevant materials for a problem or concept and let them "browse" and test ideas until they discover relationships and rules on their own. A slightly different way of fostering discovery learning—called by some "guided discovery"—is to guide the children through all the steps or conditions leading up to a conclusion but let them come up with the actual rule themselves. Advocates of discovery learning claim that these methods somehow "fix" the newly learned concept in memory so that it is held longer and generalizes better to new situations. Children taught by discovery methods should, in other words, perform better at Katona's comparison stage than children who were simply told the same principles. Tied to this is the notion that if children are encouraged to formulate the discovered rules for themselves, the rules are internalized to a greater degree than if they are simply accepted in the form handed down by the teacher. This should make the rules more readily available for problem solving and for understanding related mathematical topics.

A conference held in 1965 (Shulman & Keislar, 1966) brought together a number of eminent scholars to review the psychological literature on discovery learning and discuss its pros and cons. In a sense, this meeting demonstrated that discovery learning was a poorly understood term, for though it had intuitive appeal—certainly we all have a notion of what discovery is, based on our own experiences—it was extremely difficult to define for the purpose of rigorous scientific experimentation. As a result, the studies that had been done could not confidently be used to back up instructional practice, even though discovery was being advocated by some as *the* best way to learn. The suggestion emerging from the conference was that the concept of discovery should be broken down into a number of more specific and experimentally manageable subtopics. It was also proposed that analysis of results should center around a whole set of interactions—"subject matter, with type of instruction, with timing of instruction, with type of pupil, with outcome [Cronbach, 1966, p. 92]"—rather than simply comparing a discovery method with a nondiscovery method.

Of interest for the present discussion is that in a number of studies on discovery learning people in the nondiscovery conditions were given no *chance* to find the structural bases of the problems they were to solve. They were shown fixed ways of solving the problems without explanations for why these rules worked. Then they were encouraged to apply the rules to many examples in exactly the way shown them. They thus had no reason to look for structure or meaning in the problems; they were simply "following orders." Or sometimes, as in Group 1 of the Katona experiment just described, nondiscovery groups were given rules that actually interfered with noticing the structure of the problem. The experimental discovery groups, on the other hand, were given materials to help point up the structural bases of problems and were given opportunities to try out different ways of putting things together in a way that made sense. These people were being encouraged to notice the underlying mathematical structures that made problems meaningful. In some of Katona's experiments the meaningful learning group did not have to discover the underlying principle or structure for themselves. Instead they were shown the principle directly and then given a chance to study it and apply it to new problems. These students did not perform much differently from students who discovered the principle for themselves.

A study by Gagné and Brown (1961) also suggests that the content rather than the method of instruction may account for the apparent superiority of discovery teaching. In this study teenage boys were taught to obtain formulas for the sums of specific number series. A "rule-and-example" group was shown the formulas and then led through a number of examples in which they identified terms and practiced finding the numerical values of the series using the formula. "Discovery" and "guided-discovery" groups, on the other hand, were asked to derive the formula for the sum of each series and were given increasingly explicit hints to help them achieve the solution. Both the rule-and-example and the guided-discovery programs proceeded in small, carefully sequenced steps based on general principles of programmed learning. In the test situations, the groups were presented with a number series that they had not seen before and were required to derive the formula for its sum. The guided-discovery group was most successful in the novel problem-solving task, followed by the discovery group and the rule-and-example group in that order. Looking for the crucial differences between the guided-discovery and the rule-and-example learning programs, one notes that the guided-discovery program directed students' attention to ways of finding relationships between numbers in the series, whereas the rule-and-example group simply practiced applying a formula in which the relationships were already given. The two programs thus taught different content and one (guided discovery) was more useful when no formula was available, as in the test situation. Like Katona's, this finding suggests that it is not the *discovery* itself but the *principle* that is important for meaningful learning. Where the aim of teaching is retention and transfer, it may be as effective—and is usually more efficient—to demonstrate the principle directly rather than requiring students to discover it.

IMPLICATIONS OF GESTALT THINKING FOR INSTRUCTION

Having reviewed briefly the background of gestalt psychology and the demonstrations and experiments of some of its major theorists, our task is to evaluate the relevance of gestalt principles for mathematics instruction.

We saw, first, that in the case of the carpenter's apprentice and parallelogram problems, the explanations or demonstrations that provided insight into their solutions were couched in terms of the underlying mathematical or geometric structures. This suggests that a task for instruction is to present problems in ways that highlight their various interrelated components and promote insight into their underlying structures. Although Wertheimer did not spell it out in so many words, we can infer from his demonstrations that the structures that underlay productive thinking could often be defined as the structures of mathematics. It was apparent to the gestalt psychologists that drill approaches to learning stamped out the natural tendencies to organize thinking in structure-based ways.

In a sense, the human tendency to impose structure in thinking and perception—to look for "good gestalts"—provides a theoretical rationale for the kind of structure-oriented teaching methods spawned during the era of the Cambridge and Woods Hole Conferences. Gestalt psychology, like the psychology of Piaget, views the learner as an organizer of perceptions and experiences. This being the case, it is not unreasonable to want to teach portions of larger mathematical structures in a spiral curriculum. To the extent that teaching highlights the structural properties of concepts and procedures, the larger organization may be provided in time by the learner. Because gestalt psychology credits learners with more global organizing processes than simple stimulus–response learning theories would suggest they have, it gives credence to the possibility that children may be able to discover mathematical principles for themselves in the process of working with specially designed math materials.

Perhaps a fruitful way to consider the theory and experiments presented in this chapter is to think of them as pointing out the importance of building good mental representations, both of the subject matter of mathematics and of specific problem-solving tasks. In the preceding chapter, we discussed Bruner's hypothesized enactive, iconic, and symbolic modes for representing the structures of mathematics in memory. We have expanded the definition of a mental representation here to include the way the structure of a mathematical problem is conceived. How the subject matter and problem structure are mentally represented is not spelled out in gestalt writings, except for the suggestion that those structures might activate certain neurological patterns that are isomorphic to patterns previously perceived in the environment (Köhler, 1929). We assume that the particular representation an individual has of a subject matter, including relevant procedural knowledge, determines how he or she enters into problem solution. We also assume that the mode of instruction has an impact on the form of people's mental representations. In Chapter 8, we report studies that seek to map out the exact organization of subject-matter and procedural knowledge for the purpose of developing a theory of mental representation.

If problem representations are superficial, as a gestalt psychologist might expect in cases where rote instruction has dominated mathematical learning, then the thinker may be limited in the problem-solving resources he or she can bring to bear. Such a person will tend to make the same mistake as the apprentice in Wertheimer's staircase story and will find the sum of a series by a long and tedious computation. If the representation is based on the underlying mathematical structure of the problem, on the other hand, the problem solver is capable of experimenting with different ways of setting up the problem and is more apt to find a simple and elegant approach to solution. In this conceptualization, a previously taught rule or algorithmic solution need not be considered less desirable than a personally constructed original solution to a problem—indeed, to require original solutions would be to negate the cumulative efforts of centuries of mathematical thinking. Perhaps, then, the criterion of productive thinking should be whether or not the representation underlying a problem solver's actions is rich enough to make contact with the knowledge he or she has built up concerning the structure of the subject matter.

Besides the rather general notion of structure-oriented teaching, what can we offer in the way of suggestions for increasing the probability of insight into any given problem situation? Although the gestaltists attached great importance to insight in their writings, they largely failed to analyze the processes in insight and provided little concrete guidance for would-be problem solvers. The vague suggestion that insight depends on getting at the "good inner relatedness" of a problem (Wertheimer) or understanding the inherent "conflict" in a situation (Duncker) is less than adequate when our concern is to improve individuals' ability to solve problems.

We interpret Wertheimer as saying that insight comes about in part as a result of understanding the structures of mathematics. The studies on meaningful and discovery learning highlight the need for adequate representations of problems and their underlying mathematical structures. But beyond the understanding and representation of structure, there are features inherent in any problem that draw upon another aspect of mental processing: solution strategy. By definition, a true problem requires a search for possible solutions because no solution is readily apparent. This implies some sort of strategy for determining what the goal of the problem is, what information the learners have at their disposal, and what the missing information is that would allow either a known solution strategy to be applied or a new solution to be invented. This aspect of problem solving is largely ambiguous in gestalt writings, although Duncker went part way in trying to analyze the processes. Today the roles of understanding and insight in problem solving still loom large as topics for research, and studies are underway that may shed more light on these complex intellectual processes that problem solving brings into play. We explore these concepts further in Chapter 8, as we present information-processing theories of problem solving. Meanwhile, we are offered some concrete suggestions for increasing the likelihood of insight and enhancing general problem solving ability in the work of Polya, a mathematician influenced by gestalt theory and directly concerned with instruction.

Facilitating Insight

George Polya spent years showing teachers how to teach mathematics, focusing specifically on those kinds of helps and hints that guide people toward insight into problem solutions. His suggestions can be viewed as ways of facilitating

Implications of Gestalt Thinking for Instruction   149

discovery of the underlying structures, in the gestalt sense, of the problems to be solved. Polya provides a set of specific questions or steps to follow in working on a problem, each of which could be characterized as *heuristic,* a word of Greek origin meaning *serving to discover.* The value of heuristics is that they allow a would-be problem solver to proceed systematically toward insight, instead of leaving "productive thinking" to chance or to the gifted few who are quick to see problem structure on their own.

According to Polya (1945/1957), problem solving may be divided into four stages, namely, understanding the problem, devising a plan for finding the solution, carrying out the plan, and looking back to verify the procedure and check the result. These four stages, with detailed questions and hints relevant to each, are listed in Fig. 6.5. Let us examine them in the context of a specific problem:

> A stunt motorcyclist plans to ride his bike the length of a tightrope stretched from the upper left-hand corner of the back of an auditorium to the lower right-hand corner of the front of the auditorium. The dimensions of the auditorium are 100 by 60 by 30 feet. He needs to know how long a tightrope he should bring to span the distance he plans to ride. How would you help him find out?

Now, imagine a sixth grader, Brian, trying to follow Polya's steps. *What is the unknown?* "The length of the tightrope." *What are the data?* "A room 100 by 60 feet and 30 feet high." *What is the condition?* Here Brian would probably want to draw a figure and label the knowns and unknowns appropriately, as in Fig. 6.6. "The condition is that t spans the distance from point A to B of a three-dimensional rectangular figure, with dimensions w, h, and l." *Look at the unknown. Do you know a problem with the same or a similar unknown?* ... "Yes, finding the hypotenuse of a right triangle...." So there is a related problem. *Can you use it?* ... Hopefully by this time Brian has enough hints to go on to solution of the problem but, if not, Polya suggests giving increasingly explicit hints, helping him bring to mind other problems he has solved in the past and procedures he already knows for finding the length of a line segment.

What we are searching for in this example is the moment of insight. It is what occurs between the time Brian says "Gee, I have a problem here. I wonder how I can solve it" and "Oh yes, this is basically a case of finding the hypotenuse of one triangle and substituting it into another find-the-hypotenuse problem." It is this crucial step that seems to be affected by the kinds of principles gestalt psychologists speak of. It involves looking for structure, reformulating the problem, and perhaps trying to loosen up a preconceived structure so that new ways of looking at the problem present themselves. Once the structure (hypotenuse of a right triangle) is apprehended and the nature of the parts of the problem (e.g., the data) in relation to the whole (the conjunction of line segments that makes this a triangle problem) is recognized, Brian is ready to formulate a systematic plan to

UNDERSTANDING THE PROBLEM

First.
You have to *understand* the problem.

What is the unknown? What are the data? What is the condition? Is it possible to satisfy the condition? Is the condition sufficient to determine the unknown? Or is it insufficient? Or redundant? Or contradictory?

Draw a figure. Introduce suitable notation. Separate the various parts of the condition. Can you write them down?

DEVISING A PLAN

Second.
Find the connection between the data and the unknown. You may be obliged to consider auxiliary problems if an immediate connection cannot be found. You should obtain eventually a *plan* of the solution.

Have you seen it before? Or have you seen the same problem in a slightly different form?

Do you know a related problem? Do you know a theorem that could be useful?

Look at the unknown! And try to think of a familiar problem having the same or a similar unknown.

Here is a problem related to yours and solved before. Could you use it? Could you use its result? Could you use its method? Should you introduce some auxiliary element in order to make its use possible?

Could you restate the problem? Could you restate it still differently? Go back to definitions.

If you cannot solve the proposed problem try to solve first some related problem. Could you imagine a more accessible related problem? A more general problem? A more special problem? An analogous problem? Could you solve a part of the problem? Keep only a part of the condition, drop the other part; how far is the unknown then determined, how can it vary? Could you derive something useful from the data? Could you think of other data appropriate to determine the unknown? Could you change the unknown or the data, or both if necessary, so that the new unknown and the new data are nearer to each other? Did you use all the data? Did you use the whole condition? Have you taken into account all essential notions involved in the problem?

CARRYING OUT THE PLAN

Third.
Carry out your plan.

Carrying out your plan of the solution, *check each step.* Can you see clearly that the step is correct? Can you prove that it is correct?

LOOKING BACK

Fourth.
Examine the solution obtained.

Can you *check the result?* Can you check the argument? Can you derive the result differently? Can you see it at a glance? Can you use the result, or the method, for some other problem?

FIG. 6.5 Polya's stages of problem solving. (From Polya, 1957. Copyright 1957 by Princeton University Press. Reprinted by permission.)

Implications of Gestalt Thinking for Instruction   151

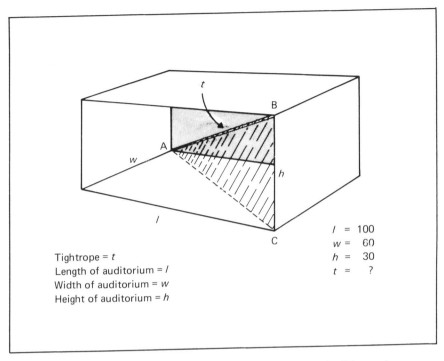

FIG. 6.6 Labeling unknowns and drawing a figure facilitates visualizing a solution procedure for the tightrope problem.

solve the problem. Then, provided he chooses appropriate algorithms or procedures, all that remains is for him to compute and check the results.

Notice how much more clearly Brian could "see" the tightrope problem when he drew a diagram of it. In a previous chapter we noted, too, that drawing a picture of an "impossible" word problem led quickly to the recognition that a direct algebraic translation would have yielded an erroneous solution. Wertheimer showed us how to simplify the staircase problem by visualizing it as complementary parts of a rectangle. Drawing diagrams and pictures seems to aid in both problem reformulation and goal analysis, processes that are heavily implicated in insight. As is true of Wertheimer's examples, many of the sample problems in Polya's books for mathematics teachers are geometry problems or problems that lend themselves easily to visual representation. Although seemingly limited in their generality, these types of problems seem to demonstrate particularly well the structural principles upon which problem solving (and, more generally, mathematics) is based. Spatial representation is in fact an integral and necessary aspect of higher mathematical thinking. Even at the elementary level, we value the manipulation of concrete materials because our intuitions tell us the visual and spatial aspects of simple mathematics are important.

With respect to the process of human problem solving, Polya (1962) has said:

> Solving problems is the specific achievement of intelligence, and intelligence is the specific gift of man. The ability to go round an obstacle, to undertake an indirect course where no direct course presents itself, raises the clever animal above the dull one, raises man far above the most clever animals, and men of talent above their fellow men [Vol. 1, p. 118].

"The ability to go round an obstacle, to undertake an indirect course where no direct course presents itself"—this brings to mind Köhler's apes who had to make a mental detour from their goal of getting a banana in order to notice that other objects about them could be employed in solving their problem. It is that detour that Polya encourages by asking "Do you know a problem with a similar unknown?" or "What else do you know that might work here?" It is that detour that facilitates insight into the problem solution. Gestaltists would explain this insight in terms of apprehending structures of problems. Thus Polya, concerned with general principles of discovery and deriving his methods from a mathematician's background, operates within a framework that we can understand in terms of the structural thinking of gestalt psychology. At the same time, he provides specific hints that appear to encourage a kind of bottom-up processing growing out of constituent problem features. Thus, we may view Polya as highlighting a necessary interaction between the whole and its parts, between the integrative and analytic aspects of problem solving.

SUMMARY

In this chapter we have looked at a theoretical position that contrasts with early learning theory and its analysis of behavior in terms of stimulus–response associations. The central thesis of gestalt psychology is that thinking and perception are dominated by an innate tendency to apprehend structure. This being the case, the experience of perceiving or thinking achieves an organization that is more than the sum of objectively identifiable individual elements or stimuli. The organization of experience is controlled by the tendency to seek good gestalts, closure, or psychological equilibrium. In demonstrations of the dominance of the whole over the parts in mathematical problems, Wertheimer tried to show that apprehending the underlying structures, interpretable as mathematical structures, led to productive thinking and elegant problem solving. This was because, by gaining insight into problem structures, the problem solver understood the relevance and functions of problem components and of known solution procedures, suggesting paths to solution.

The *process* of achieving insight, although largely unexplored by gestalt psychologists, was analyzed in part by Duncker. Duncker drew a distinction between processing from above, starting with analysis of goals and problem reformulation, and processing from below, starting with the analysis of problem materials. The same distinction is made in current analyses of cognitive performance, although in practice the interaction between the two types of processes, rather than their differences, may be more amenable to study.

Attempting to develop a gestalt theory of learning, Katona suggested that meaningful learning, like problem solving, depended on a presentation that either made explicit or allowed the learner to discover the underlying mathematical structure. If the principles underlying the content of learning and problem solving were understood, solutions could be reconstructed, extended, and remembered. In contrast, rote learning appeared to limit the learner's ability to remember and generalize new learning. These contentions were similar to those found in the more recent literature on discovery learning. A review of that literature suggested that discovery is not a unitary concept and is hard to explicate experimentally. Studies purporting to test the efficacy of the method confounded discovery with the opportunity to apprehend underlying structures of material to be learned.

The gestalt view of problem solving is that insight grows out of an understanding of the problem as a whole and of the relation of the parts to the whole. Influenced by gestalt theory, Polya has developed hints that encourage the problem solver to reconsider the goals of the problem, search memory for similar problems solved before, and analyze the materials or givens of the problem. These hints may be helpful in promoting the problem reformulation and goal analysis that appear to facilitate the emergence of insight.

Gestalt psychologists have given us intuitively appealing demonstrations of the organization of thinking and perception, and they foreshadowed many of the concerns of today's cognitive psychologists. Of themselves, however, the demonstrations are not clear with respect to many of the processes we hope to influence through instruction. The nature of mental representations and the processes by which problems are formulated and solution strategies chosen are issues that require clarification. These topics are receiving further treatment in current analyses of problem-solving behavior, and we return to them in Chapter 8.

REFERENCES

Bell, E. T. *Men of mathematics.* New York: Simon and Schuster, 1977.
Cronbach, L. J. The logic of experiments on discovery. In L. S. Shulman & E. R. Keislar (Eds.). *Learning by discovery: A critical appraisal.* Chicago: Rand McNally, 1966.
Duncker, K. On problem-solving. *Psychological Monographs,* 1945, *58*(270), 1–112.
Gagné, R. M., & Brown, L. T. Some factors in the programming of conceptual learning. *Journal of Experimental Psychology,* 1961, *62*(4), 313–321.

Katona, G. *Organizing and memorizing: Studies in the psychology of learning and teaching.* New York: Hafner, 1967. (Originally published 1940.)

Köhler, W. *The mentality of apes.* New York: Harcourt, Brace & World, 1925.

Köhler, W. *Gestalt psychology.* New York: Liveright, 1929.

Kreutzer, M. A., Leonard, Sister C., & Flavell, J. H. An interview study of children's knowledge about memory. *Monographs of the Society for Research in Child Development,* 1975, *40*(1, Serial No. 159).

Luchins, A. S., & Luchins, E. H. *Wertheimer's seminars revisited: Problem solving and thinking* (Vol. II). Albany, N.Y.: Faculty-Student Association, State University of New York, 1970.

Polya, G. *How to solve it* (2nd ed.). New York: Doubleday Anchor Books, 1957. (Originally published 1945.)

Polya, G. *Mathematical discovery: On understanding, learning, and teaching problem solving* (Vol. 1). New York: Wiley & Sons, 1962.

Shulman, L. S., & Keislar, E. R. (Eds.). *Learning by discovery: A critical appraisal.* Chicago: Rand McNally, 1966.

Wertheimer, M. Untersuchung zur Lehre von der Gestalt, II. *Psychologische Forshung,* 1923, *4,* 301–350. Translated and condensed as "Laws of organization in perceptual forms" in W. D. Ellis (Ed.), *A source book of gestalt psychology.* New York: Harcourt, Brace & World, 1938.

Wertheimer, M. *Productive thinking* (enlarged ed.). New York: Harper & Row, 1959. (Originally published 1945.)

7 Piaget and the Development of Cognitive Structures

We turn now to the work of Jean Piaget and a somewhat different view of cognitive structures than that conveyed by the gestalt movement. Gestalt psychologists, as we have seen, focused particularly on the immediate way in which structures of problems or of subject matters were perceived, as if entire structures were taken in "at a glance." Because of its emphasis on the immediacy of insight and the relatively complete understanding that usually ensued, gestalt psychology seemed unconcerned with how knowledge of relationships built up to the point where such insight and recognition were possible. Nor did the gestaltists seem concerned with how, over extended periods of time, people's capacities for recognition and insight might change. In contrast, Piaget was explicitly concerned with the process and development of thinking. He also believed that the fundamental characteristics of human thinking could be understood in terms of the logical propositions and relationships that human behavior expressed. Both his interest in logic and his concern with how thinking is modified during growth and experience helped to shape his definition of cognitive structure.

Piaget is best known for his extensive studies of the development of children's thinking. Most discussions of his work particularly emphasize the idea of stages of development, and many summaries of Piaget's theory outline the sequence of stages that he proposed in the course of his research. We outline these stages and consider what they imply for teaching mathematics to children. However, we begin our discussion of Piaget with an attempt to understand his view of the role of structure in thinking. Later, we review some of the major alternative interpretations of intellectual development that have been proposed by "neo-Piagetians" and some anti-Piagetians. We also consider and evaluate the various instructional implications that are often drawn from Piaget's work, particularly those related to mathematics.

THINKING AS STRUCTURING

Piaget began his scholarly career as a biologist, and this orientation has permeated virtually all his work in psychology as well. As a biologist, he was interested in the physical structures that characterized organisms. He noted that those structures underwent a gradual development over generations so as to make the organisms better adapted to their environments. As a psychologist, Piaget was interested in cognitive structures—the structures of thinking. Although cognitive structures could not be observed directly, as physical ones could, Piaget tried to reveal the thought processes of children through a technique of activity-based interviewing. His style of interviewing, displayed in sample protocols throughout this chapter, provides an important alternative method of investigation for psychologists and educators interested in probing the nature of individuals' thinking and understanding. Although Piaget's research is based exclusively on clinical interviewing, the greatest value of the method to psychology probably lies in its judicious combination with other more experimental methods, much as combining protocol analysis with more quantitative research strategies is proving profitable in information-processing research.

Much of Piaget's work was premised on the notion that individuals recapitulate, in the course of their development, the intellectual history of the human species. Piaget thought it possible, therefore, to understand the development of the species' intellectual capacities by studying the intellectual development of individuals as they grew into adults. As Piaget's work progressed, he became increasingly convinced that certain basic structures of thinking, which could be defined logically and mathematically, were inherent for human beings. By inherent, Piaget did not mean that people were born with these structures fully formed or that children raised apart from normal human relationships would develop them. He meant, rather, that all humans would develop certain structures of thinking as long as they maintained a normal interaction with both the social and physical environment. The idea was that people were biologically constructed to interact in certain ways with their environment. In the course of this interaction a sequence of complex structures of thinking would emerge.

Although the development of children's thinking could be studied in many subject areas, Piaget's most extensive work was on the development of logical and mathematical concepts. He studied, in both children and adolescents, the growth of logical classification systems and the concepts of number, geometry, space, time, movement, and speed. These topics were chosen for intensive study because they clearly involved the use of certain basic logical structures. Along with earlier philosophers who studied epistemology—the science of knowledge—Piaget believed these structures were the basis of thinking and reasoning, particularly of a scientific kind. We can best convey Piaget's notion of structures by considering examples from his research. We begin with some experiments on children's conceptions of geometry (Piaget, Inhelder, & Szeminska, 1948/1960).

The Angles of a Triangle

A child is shown a triangle, and it is cut, as shown in Fig. 7.1a, so that the three angles can be picked up and handled separately. Two of the angles are arranged side by side (Fig. 7.1b). The child is asked to predict what they will look like when the third angle is added. After the prediction, the third angle is added and the result is a half-moon (Fig. 7.1c); that is, the angles sum to 180 degrees, or form the equivalent of a straight line. Will this be the case for the sum of the angles of all triangles?

The point of interest is not what the child actually knows about the angles of a triangle at the beginning of the experiment but how the child thinks about the problem during the experiment, and whether by the end he or she becomes quite convinced that the three angles put together will always form a straight line (i.e., that they will always sum to 180 degrees). To find out how the child is thinking, the experimenter presents the problem several times using triangles of different shapes and sizes. Sometimes, after the angles have been arranged in a line, the experimenter rearranges their order and asks the child to predict whether they will still fit along the line. This is a way of checking whether the child understands that the order of angles does not affect their sum. At other times the experimenter tries to fool the child by taking the third angle from a different triangle and asking how it will look alongside the two angles from the first triangle. Or, late in the experiment, after the child appears convinced that the angles will always sum to 180 degrees, the experimenter introduces a set of angles summing to more or less than 180 degrees and asks how this could be (see Fig. 7.1d). A child whose understanding of the nature of a triangle is secure is likely to guess that the pieces were taken from more than one triangle. What happens under these various conditions, and how are Piaget's structures of thinking revealed?

In the following paragraphs we discuss samples of verbal behavior elicited by Piaget while studying children's conceptions of geometry. The sample protocols are shown in accompanying figures. The protocols typify the kind of dialogue between experimenter and subject that has characterized Piaget's clinical style of research over the years. Individual children are presented a problem or situation and are asked to verbalize their thinking as they proceed; the experimenter questions and probes each child's thinking, pursuing thoughts in the order and depth suggested by children's verbalizations. Questioning is led, in part, by the moment-to-moment responses of the children.

158   7. PIAGET AND THE DEVELOPMENT OF COGNITIVE STRUCTURES

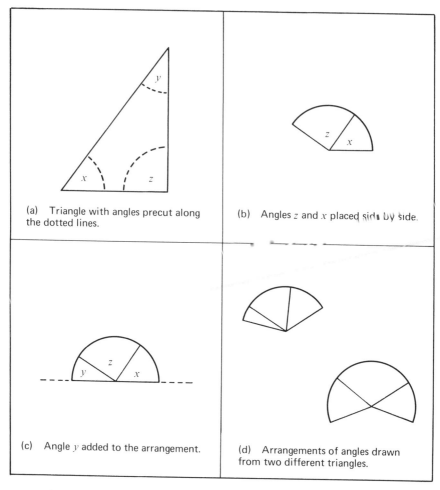

FIG. 7.1 The angles-of-a-triangle problem. Some children discover that the angles of a triangle always sum to 180°.

Some children, usually the 5- and 6-year-olds, are quite unable to treat the angles-of-a-triangle problem as anything but a string of unrelated guesses. They make predictions but apparently without system and without justification. Figure 7.2 gives an example of such a child, Ful. At first, Ful cannot make a prediction at all (Fig. 7.2a), although he can recognize and name a half-moon after the angle pieces are put together. When asked for a prediction on a second triangle, he appears to have generalized from this first experience, because he predicts the angles will form a half-moon again (b). But we soon see that he has not really understood the principle: He predicts a sum of less than 180 degrees for the angles of the next triangle (c) and gives no reason for his prediction (d). He also

EXPERIMENTER	FUL
(a) Presents right angle triangle with precut angles.	Puts together three precut angles taken from the right angle triangle and sees that they form a semi circle although he failed to predict it.
(b) Presents another triangle. *What if I put these two corners together and then put the third corner alongside, what will these make?*	*They'll make a half-moon as well.*
(c) Shows another right angle triangle, much larger. *What if I do the same with the corners of that triangle?*	Indicates a figure less than 180°.
(d) *Why will it be less?*	Doesn't reply. Experiments with the pieces. *Oh, it's the same!*
(e) *Now what if I take this corner on this side and put it over on that side instead of this one and put this one over here?* Indicates an exchange of pieces as he speaks.	*It'll be bigger.*
(f) *Are you sure?*	*No, it'll be half-moon.*
(g) *Why?*	*No, it will be more.*
(h) Presents a trapezium (four-sided figure). *And what if we put the corners of this figure together?*	*They'll make a whole circle because there are several large ones.*
(i) Presents a small equilateral triangle. *And the three corners of this one?*	*They'll make a half-moon and a little bit.*
(j) *Why?*	*These two little corners make a half-moon.* Experiments. *Oh, no! It's a half-moon.*
(k) *And if I change these two?* Indicates exchange as before.	*A half-moon leaning over... I don't know.*

FIG. 7.2 Experimenter interviewing child working on the angles-of-a-triangle problem. Ful, 5½ years old, demonstrates an incomplete understanding of the problem. (Adapted from Piaget, Inhelder, & Szeminska, 1960. Copyright 1960 by Routledge & Kegan Paul, published in the U.S. by Basic Books, Inc.)

thinks that rearranging the pieces will make the set larger (e). He is easily influenced by the experimenter, for when asked if he is sure about the rearranged pieces (f), he reverts to the half-moon prediction. Then, in response to the experimenter's "Why?," he says it will be more (g). Apparently, Ful has no strong idea of his own and is just following what he thinks are the experimenter's clues. When a four-sided figure, the trapezium, is presented, Ful makes a correct prediction (h), but we cannot tell whether he recognizes that the angles of four-sided figures will always sum to 360 degrees. Probably he does not, for on the next triangle (i, j) he is estimating again, apparently from perceptual features; this time he thinks two of the corners will make a half-moon by themselves, and the third will make "a little bit" more. And once again (k) he thinks reversing position might change quantity, but he is not sure.

Compare Ful's performance with that of another child, Jeq, 5 years older. His responses appear in Fig. 7.3. Although Jeq has some trouble with verbalization, he nevertheless makes it clear (b) that he is looking at the three angles in relation to one another and thinking about how the angles and directions of the lines in the figure are related. He sees that if the lines are to meet, it is necessary that the angles sum to only 180 degrees. If they sum to more than that, there must be four or more angles; that is, the figure must be some shape other than a triangle (b, d). Reluctant at first to predict the sum of angles, Jeq proceeds to make a prediction before he has finished assembling the cut angles (c). Then he gets the idea that the angles will always form a semicircle (c) and sets out to test (d) what he is quite obviously treating as a hypothesis.

Jeq indicates at several points that he is considering the angles in relation to each other: "One big and two thin" (d), "These two angles aren't quite right angles, they're made up by the one at the top" (e). In other words, he sees the triangle not just as a given perceptual form but as an interrelated system of lines and angles. He recognizes, too, that modifying any one part of the system will require a compensating modification in some other part. Thus, the enlargement of one of the angles must be offset by a reduction in one or more of the other angles if the system is to remain intact, that is, if it is to remain a triangle.

A comparison of the performances of Jeq and Ful highlights Piaget's use of the term *structure*. Jeq responds to the triangle in terms of interrelated features—lines and angles—and sees that an action in any part of the triangle will change the whole figure. In other words, Jeq understands the figure in terms of relationships he creates through his thought processes; he conceives the figure as a whole, features of which can be thought of independently. Ful, on the other hand, is not able to think of all of these relationships simultaneously. He does sometimes try to consider more than a single feature at a time. But he is not very good at it, apparently because he does not see any connection between the lines and the angles of the triangle. He is thus reduced to guessing and perceptually approximating. His analysis of the problem is unsystematic, and he lacks confidence in his judgments.

EXPERIMENTER	JEQ
(a) Presents right angle triangle with precut angles.	Assembles three angles. *A semi circle.*
(b) *Why?*	*I can't tell you, it's because the angles are smaller than they would be if they were right angles. . . . You'd need four; three angles wouldn't meet* (if they were greater than 180°).
(c) Presents another triangle. *And that one?*	*I can't say in advance.* Prepares to assemble the angles. *It's another semi circle. I'm beginning to think it's always a semi circle. With three angles you can't have a full circle because you'd need angles bigger than a right angle; you'd need a further stroke* (i.e., side).
(d) Presents an isosceles triangle with a very obtuse angle.	*One big and two thin; a semi circle again! I want to try again with that one* (an even more obtuse-angled triangle). *Yes, still the same because there's an enormous angle at the top.*
(e) Presents very pointed isosceles triangle. *And this one?*	*A semi circle again! These two angles aren't quite right angles, they're made up by the one at the top. Two right angles would make a semi circle?*
(f) *And if we made it so long that it reached the cellar?*	*It's always the same if the lines are quite straight.*

FIG. 7.3 Protocol of an 11-year-old child, Jeq, successfully attempting to understand the angles-of-a-triangle problem. (Adapted from Piaget, Inhelder, & Szeminska, 1960. Copyright 1960 by Routledge & Kegan Paul, published in the U.S. by Basic Books, Inc.)

Another difference between the two performances is that Jeq is quite clearly able to make predictions and test his ideas, although we have no evidence that his tests are fully systematic and exhaustive of all possibilities. By contrast, Ful does not formulate or test a general hypothesis that he thinks will apply to all triangles. Rather, his actions seem to respond solely to the experimenter's cues. Thus building structure in the Piagetian sense appears to involve constructing relationships such that change in any part of the system affects the whole system. Further, the more advanced forms of structuring lead to hypotheses about general relationships, thus freeing thought from the immediate stimulus at hand.

Matching Spatial Orientation in a Coordinate System

Let us consider another example. The task is deceptively simple: the child is shown a rectangle with a dot in it and another identical rectangle with no dot. The situation is depicted in Fig. 7.4a. The child is then asked to place a dot in the empty rectangle so that it matches exactly the position of the dot in the first rectangle. Imagine for a moment how an adult might try doing the task. Perhaps one's first inclination would be to estimate visually; no doubt one could match the position quite well by this method. But the experimenter in this study does not accept estimation. An exact match of position is required. How could the position of the dot be located exactly?

Here is what children of different ages do to solve the dot problem: The youngest children (4-5 years) are unable to move beyond visual estimation. When it is suggested that they measure using rulers and straight edges, they either use them randomly or reject them altogether. They simply do not see the problem as one in which measurement is either possible or helpful—although some are glad enough to oblige the experimenter by appearing to measure if that is what is requested. Further, during this period of development, even visual estimation is very poor. Typically children can place the dot at the same height as in the other rectangle but are unable to take horizontal position into account at the same time, as shown in Fig. 7.4b. Within a few years, children's visual estimation improves, and they can place the dot in approximately the right position in both dimensions. But when asked to measure, children of 6 or 7 years seem unable to take both dimensions into account at once. They measure height and forget width, or vice versa. When they are estimating visually, they can coordinate height and width; but they are unable to coordinate the two dimensions in a rigorously quantified system (i.e., one that involves measurement).

In time, most children overcome this problem, but their first efforts are of a special kind. What they do is to lay the ruler along the line from the corner of the rectangle to the point in the first figure, as shown in Fig. 7.4c. Then they move the ruler over to the blank rectangle, trying all the while to maintain the slope of the ruler. Of course this introduces new kinds of inaccuracies, but at least it is an attempt to take both dimensions into account, even if it means reducing them to a single one. Only later, toward the age of 8 or 9 years, do most of Piaget's subjects recognize that to obtain a truly accurate placement of the second dot they need to make two separate measurements and them combine them. The subjects described by Piaget do this by measuring height and width separately and placing the dot at the intersection, as shown in Fig. 7.4d.

With this example, we can expand further our understanding of structure as conceived by Piaget. Here again, as with the triangles, we can see that accurate performance on even a simple geometric task involves understanding how several separate parts relate to each other. To place the dot correctly the child must act with respect to the whole, not just one dimension or the other. Further, using

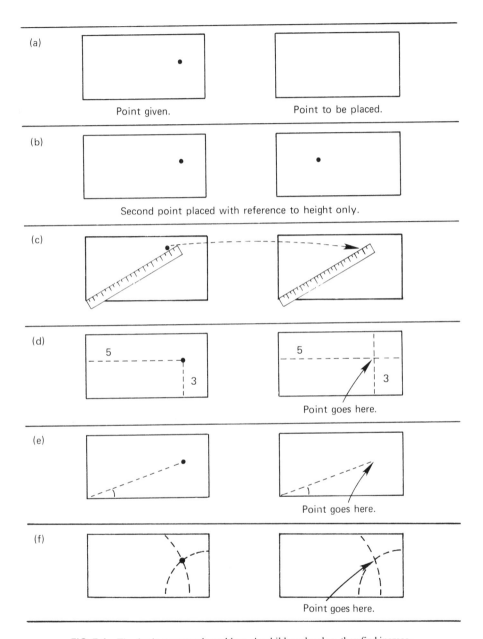

FIG. 7.4 The dot-in-a-rectangle problem. As children develop, they find increasingly sophisticated ways of figuring out where to place the dot in the empty rectangle.

164 7. PIAGET AND THE DEVELOPMENT OF COGNITIVE STRUCTURES

rectangular coordinates, as opposed to simply estimating visually, requires imagining lines in an empty space and then partitioning those lines into equal units (the essence of measurement). Finally, with analysis and partitioning completed, all the actions must be coordinated so that the dot can be positioned. This problem illustrates the point that structure is not provided from the outside; children do not just look at the rectangle and know immediately where to place the dot. They have to engage in some intellectual activity to untangle the various aspects of the problem. In the present case the activity involves mentally constructing and using a rectangular coordinate system, which is the basis for much graphing and thus important not only in geometry but also in algebra, calculus, and other branches of mathematics. Children who accurately place the dot by this method behave as if there were a grid of horizontal and vertical lines, equally spaced, laid over the rectangle. They count up and across the imaginary grid to match the dot's position in one rectangle with a point in the empty rectangle. The grid is not physically there; the children must construct it. They can respond only by actively operating upon the material presented to them.

Piaget does not mention it in this context, but there are at least two other ways of accurately placing the dot. One method is to measure the distance from the dot to one corner of the rectangle and also to measure the angle created by the base of the rectangle and the line of measurement (Fig. 7.4e). These two measurements repeated on the empty rectangle will accurately determine the position of the dot. Another method is to measure the distance from the dot to each of two corners of the rectangle, construct arcs, and place the dot at the intersection of the arcs (Fig. 7.4f). Both of these methods require constructions by the child and require coordination of two separate measurements. They would thus illustrate Piaget's point about active structuring as well as the rectangular-coordinate-system method does. Yet Piaget's experiments in which rulers were provided but no protractors or compasses, allowed only the coordinate-system method to emerge. Because constraints of this kind are typical in Piaget's experiments, it is important not to take any particular study as evidence of a universal strategy of thought.

To summarize, actions—either mental (for it is not necessary *actually* to draw in all the lines) or physical (because they *could* be drawn in)—appear necessary for thinking and are part of Piaget's definition of structure. It is the nature of thinking to operate on the material presented rather than just accepting it. For Piaget, learning as well as performing mathematics is a matter of active thinking and of operating on the environment, not of passively noting or even memorizing what is presented.

With this characterization in mind it is useful to distinguish between gestalt and Piagetian definitions of structure, although the definitions are not mutually exclusive. The Piagetian examples used so far are from geometry, as were the problems of Wertheimer and Polya in Chapter 6. In both sets of examples, "structure" refers to some representation of the subject matter or problem situation involving the relations of parts to the whole. In the gestalt sense, however, structure is something that is perceived because of a tendency to recognize particular organized wholes, or "good gestalts," in the environment. Piagetian structure, on the other hand, is something actively constructed by the human organism. The understanding that emerges from this activity bears a direct relation to what one might call the objective subject-matter structure, but it is really a personal creation, with consequent variations and fluctuations with time. As Piaget (1970) has put it, gestalt psychology deals with a structur*ed* system; Piagetian psychology deals with a structur*ing* system.

THE DEVELOPMENT OF PIAGETIAN STRUCTURES

Piaget's structuring system is dynamic, flexible, and capable of changing over time. Not surprisingly, then, his conception of structure is tied to a developmental theory of human intellect. It is to this theory we now turn, keeping in mind the Piagetian notion of active structuring as it is brought to bear on the performance of intellectual tasks.

The protocols from Piaget's experiments show that children become increasingly more sophisticated in their thinking as they become older. According to Piaget's interpretation, they take more characteristics of any situation into account and recognize how transformations in one part of an organized system will affect others. They are also able to carry out several operations, to recombine them, and to reverse their own operations mentally. This correlation of age with increasingly sophisticated thinking is central to Piaget's theory of intelligence and mental development. The essence of this theory is that, as people grow older, they do not just acquire more knowledge, they develop new, more complex cognitive structures.

Central to Piaget's developmental distinctions, as they relate to mathematical and other thinking, is the concept of an *operation*. We saw in the dot-in-a-rectangle task that successful solution required dealing actively with the materials presented. An operation is a special kind of mental action, special in that it can be reversed. It can be undone by performing another action. For example, a set of five blocks can be transformed by removing three blocks, but those blocks may be restored mentally if one wishes to deal with the original quantity. Thus, an operation can transform a system (subtracting 3 from 5 leaves 2), but another operation will restore the system to its original state (adding 3 to 2 yields 5). This possibility of making and undoing transformations, also referred to as *reversibility*, is a characteristic of structures of operational thinking. But it takes some time for operations, and operational thinking, to develop. In Piaget's theory the presence or absence of certain operations is a defining feature of the stages of development through which the individual passes on the way to intellectual maturity.

Quite young children, according to Piaget's research, do not think operationally at all. They can act upon things in the environment, but once an action has been performed, they are unable to keep in mind the way things looked before. Thus, they cannot mentally reverse their actions; in the technical language of Piaget, they have not yet achieved reversibility. Instead, Piaget characterizes children during this early stage of intellectual development as heavily influenced by the sensory and perceptual features of the events that surround them. The way things are presented to them are accepted as the way things are. Mental transformations, so characteristic of older children's and adult's thinking, cannot be performed, allowing perceptual givens to dominate thinking to a much greater degree.

This predominance of perceptual ways of thinking and the inability to think reversibly are illustrated by Piaget in his well-known studies of conservation and classification. In a number conservation experiment (Piaget, 1941/1952, p. 49), a boy is shown a row of flowers and a row of vases, lined up in one-to-one correspondence (see Fig 7 5a). In this arrangement the boy can see easily that there are the same number of flowers as there are vases. But next, the row of flowers is spread out, as the child watches. In this arrangement the flowers are no longer visually matched one for one to the vases, but none have been added or taken away (Fig. 7.5b). Despite having witnessed the transformation, and in some versions of the experiment actually performing it himself, the very young child no longer views the two sets of objects as being equal in number. According to Piaget, this is because once the flowers are spread out, and the one-to-one match destroyed, he cannot imagine them back in the original position. If someone puts them back, or asks him to, there is no problem. He will then state that the sets are equal again, but he will not see—despite repeated spreading and rematching—that the quantities do not change just because the spatial arrangement is transformed. Similar difficulties are encountered with conserving other kinds of quantities. For example, young children believe that the amount of liquid changes when water is poured from a low, flattish container to a tall narrow one, and they believe that the amount of clay changes when a ball of clay is rolled out into a snake-like form.

In the typical classification experiment (Piaget, 1941/1952, p. 165), a young girl is shown a collection of two white beads and seven brown ones, all made of wood. She agrees they are all wooden, and she may even count the full set. She compares the quantity of brown and white beads and easily determines that there are more brown ones. But next the experimenter asks whether there are more brown or more wooden beads. The girl's answer: more brown ones. According to Piaget, the perceptual dominance of the large number of brown beads interferes

Piagetian Structures   167

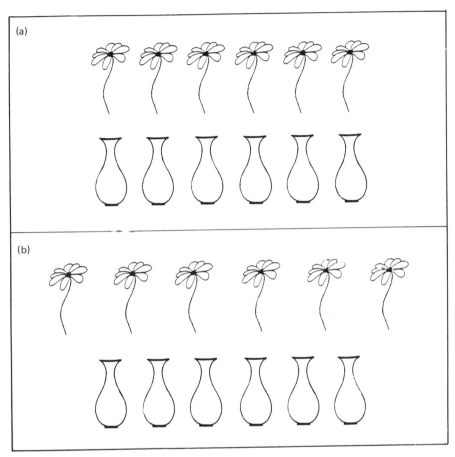

FIG. 7.5 A number conservation task. When the one-to-one match is destroyed (b), very young children judge that there are more flowers than vases because the row of flowers looks longer.

with the child's ability to take into account the fact that all of them are wooden. She compares brown with white again rather than brown with wooden. In other words, she seems incapable of comparing a subset with its own superset.

Dependence on perceptual features of objects or arrays and the inability to think reversibly are characteristic of what Piaget has labeled *preoperational* thinking. Preoperational thinking is presented in his work as typical of children of age 2 years up to about 6 or 7 years. After that, at just about the age when children are well launched in school, they enter what Piaget has termed the state of *concrete operations*. At this stage, according to Piaget, the children can think operationally: They can imagine undoing as well as doing transformations; they

168   7. PIAGET AND THE DEVELOPMENT OF COGNITIVE STRUCTURES

can think in terms of more than one dimension at a time. And as they apply these enlarged capacities for logical reasoning to more and more ideas, their mathematical and scientific conceptions move closer and closer to those of adults.

The examples discussed earlier in this chapter were largely of children whose forms of thinking would be described as concrete operational. Recognizing the structural properties of triangles and using coordinate systems to organize space systematically both require operational thinking. Further, children capable of concrete operational thinking recognize easily and with conviction that changing the spatial arrangement of a set of objects does not change its number. They recognize, likewise, that the flowers and vases presented in the conservation task remain equal in number, that the amount of liquid stays the same when poured into a new container, and that there is the same weight of clay in the snake as in the ball. In the classification experiment such children are able to consider the color and material of the beads simultaneously while still recognizing them as independent features. As a result they are able to compare the superset of wooden beads with the subset of brown ones and to declare, as adults would, that there are obviously more wooden than brown beads.

In Piaget's theory, the achievement of the stage of concrete operations marks a turning point in children's intellectual development. Particularly in mathematics far more sophisticated behaviors are possible with respect to quantity and spatial reasoning. In fact, many have argued that prior to the onset of operational thinking, any formal attempt to teach arithmetic and geometry will produce only limited understanding and limited ability to generalize and reason on one's own. But this is an argument that we can only assess in light of some of the major critiques and alternative interpretations of Piaget's findings that have been put forward in recent years. Before considering this critical literature, however, it is important to outline briefly the remainder of Piaget's stage theory.

The achievement of concrete operational thinking is great, but it is not the maximum that can be expected. According to Piaget, there is a stage of intellectual development beyond concrete operations, in which people are able to reason hypothetically and to take into account all logical possibilities. Called the period of *formal operations,* this stage typically develops with the onset of adolescence, and it involves the kind of thinking characteristic of the most advanced forms of mathematical and scientific reasoning. The following experiment exemplifies the changes that operational thinking is thought to undergo in Piagetian theory as an individual enters the stage of formal operations.

Separating Variables Experimentally: The Bending Rods Experiment

In this experiment (Inhelder & Piaget, 1958, p. 46), the child is given the task of predicting the conditions under which a rod will bend enough so that one end will touch the water in a basin. The rods vary in material (steel, brass, etc.), length, thickness, and cross-sectional form (round, square, rectangular). Three different weights can be screwed to the ends of the rods. The rods can be clamped to the edge of the water basin and can be shortened or lengthened according to where the clamp is tightened. The task is represented schematically in Fig. 7.6.

As in the Piagetian studies described earlier, the experimenter is interested not in whether the child already knows which characteristics affect the rods' flexibility but in how *the child goes about determining* the answers to such questions. The task requires considering each variable separately, and this means finding a way of holding all other variables constant while testing the effect of the one that is being considered. We begin with the case of a girl whom Inhelder and Piaget

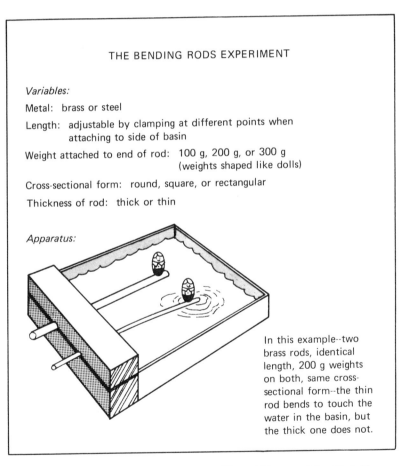

FIG. 7.6 A task requiring the ability to consider variables separately while holding all other variables constant. Children in the stage of formal operations can imagine the full range of possible combinations and construct experiments to test their hypotheses systematically.

170   7. PIAGET AND THE DEVELOPMENT OF COGNITIVE STRUCTURES

presented as an example of a formal operational thinker. The protocol appears in Fig. 7.7.

Dei begins by listing the separate variables that might affect flexibility. She then takes up one variable after another and demonstrates its effect. She does this by allowing only one feature at a time to vary. Lest her perfect performance be attributable to chance, the experimenter checks Dei's understanding by asking her to use rods that vary in thickness to determine the effect of shape (d). Dei rejects this pair of rods and explains why.

Dei's performance is so confident and so logically perfect that it is almost hard to imagine people responding in any other way. But younger children have considerable difficulty and do not succeed in demonstrating the effects of each of

EXPERIMENTER	DEI
(a) **Allows several experimental trials.** *Tell me first what factors are at work here.*	Weight, material, the length of the rod, perhaps the form.
(b) *Can you prove your hypotheses?*	Compares the 200 g and 300 g weights on the same steel rod. *You see, the role of weight is demonstrated. For the materials I don't know.*
(c) *Take these steel ones and these copper ones.*	I think I have to take two rods with the same form. Then to demonstrate the role of the metal I compare these two (steel and brass, square, 50 cm long and 16 mm² cross section with 300 g on each) or these two here (steel and brass, square, round, 50 and 22 cm by 16 mm²). For length I shorten that one (50 cm brought down to 22). To demonstrate the role of form, I can compare these two (round brass and square brass, 50 cm and 16 mm² for each).
(d) *Can the same thing be proved with these two?* **Indicates brass rods, round and square, 50 cm long and 16 and 7 mm² cross section.**	No, because that one (7 mm²) is much narrower.
(e) *And the width?*	I can compare these two (round, brass, 50 cm long with 16 and 7 mm² cross section).

FIG. 7.7   Protocol from the bending rods experiment (Dei, 16 years, 10 months). (Adapted from Inhelder & Piaget [translated by Parsons & Milgram] 1958. Copyright 1958 by Basic Books, Inc. Reprinted by permission.)

EXPERIMENTER	BAU
(a)	Experiments with the rods. *Some of them bend more than others because they are lighter.* Points to the thinnest. *And the others are heavier.*
(b) *Show me that a light one can bend more than a heavy one.* Gives him a short rod, a long thin one, and a short thin one.	Places 200 g on the long thin rod and 200 g on the short thick one, without noticing the fact that the thin rod he has chosen is also the longest. *You see.*
(c) *Show me that the long one bends more than the short.*	Again puts 200 g on the same two rods. This time the result is supposed to demonstrate the role of length.
(d) *If I take away the long one, can you compare again to find out whether it's the lightest rod that bends more?*	*Yes, this one and that one.* Points to the two short rods, one thick, one thin.
(e) *Which is better, to compare these two or to compare the way you did before?*	*These two* (long and thin, short and thick).
(f) *Why?*	*They are more different.*

FIG. 7.8 Protocol from the bending rods experiment (Bau, 9 years, 2 months). (Adapted from Inhelder & Piaget [translated by Parsons & Milgram] 1958. Copyright 1958 by Basic Books, Inc. Reprinted by permission.)

the variables separately. Consider next a 9-year-old who has no conception of the "all other things being equal" notion in scientific proof (Fig. 7.8). Later we examine the responses of some children who are intermediate in their development.

Bau uses the terms *lighter* and *heavier* to refer to thinner and thicker rods respectively; the experimenter, who is interested in Bau's logical reasoning and not the terms he uses, simply adopts Bau's language. With this in mind, it is easy to see that Bau's problems are not simply linguistic. He is quite willing to draw conclusions about weight (thickness) when the rods being compared also differ in length. Conversely, he draws conclusions about length when the rods also differ in thickness (because he is using the very same pair of rods). Under pressure he is willing to follow the experimenter; thus, when only two rods are available, he will use them to compare for the effects of weight (d). Although we recognize this as the logically correct comparison, Bau's next response (e) makes it clear that he considers it a less valuable one. His final words clinch the case. He is looking not to test for the effects of the variables separately but to create situations as different as possible.

7. PIAGET AND THE DEVELOPMENT OF COGNITIVE STRUCTURES

Bau is not totally unsystematic. He is able to observe and report what he sees accurately. Moreover he shows that he understands "multiplicative classification"—an achievement of concrete operational thinking—because he knows that rods varying in two dimensions are "more different" than rods that vary in only one (f). But as the number of variables in the problem increases, a more powerful logical system is needed, one that is able to construct mentally all the possible combinations, hypothesize effects of one or more of these combinations, and design appropriate demonstrations or experiments to test the hypotheses. These various steps require at least an intuitive understanding of propositional logic, a form of thinking that characterizes formal but not concrete operations.

Some intermediate cases (Fig. 7.9 and 7.10) now should complete the picture of how thinking changes between Bau's stage of development and Dei's. Dur has the variables separate in his mind, but he does not quite figure out how to separate them spontaneously. His first attempt (Fig. 7.9b) uses rods that vary in both weight and thickness, although he does hold material and length constant. Pressed by the experimenter, he makes a correction that yields a properly controlled comparison (c). But we cannot see from this sample of his performance how certain he is of his method or whether he will perform logically on the next try. Kra is similar. He controls for weight and length in his first try (Fig. 7.10a and b) but allows material, shape, and thickness all to vary. Pressed by the experimenter for a rigorous proof, he is able to construct a good demonstration,

EXPERIMENTER	DUR
(a)	*There are flat ones, wider ones, and thinner ones, and longer ones. If they are both long and thin, they bend still more.*
(b) *Could you show me that a thin rod bends more than a wide one?*	Puts 100 g on the round steel rod (50 cm long and 16 mm^2 cross section) and 200 g on the round steel rod (50 cm long and 10 mm^2 cross section). *That one bends more* (10 mm^2 cross section).
(c) *I would like you to show me only that the thin one bends more than the wide. Is that way right?*	Takes off the 100 g weight and puts 200 g on the 16 mm^2 rod. *You see, this is the right way.*

FIG. 7.9 Protocol from the bending rods experiment (Dur, 11 years, 10 months). (Adapted from Inhelder & Piaget [translated by Parsons & Milgram] 1958. Copyright 1958 by Basic Books, Inc. Reprinted by permission).

EXPERIMENTER	KRA
(a) *Can you show me that a wide one bends less than the narrow?*	Puts 200 g on the round steel bar (50 cm long and 10 mm² cross-section) and 200 g on the square brass rod (50 cm long and 16 mm² cross section). *This one goes down more.* Points to the thin steel rod.
(b) *Why?*	*It is round, more flexible, the steel is less heavy, it is round and narrower.*
(c) *Fine, but I would like a rigorous proof that it's because it is narrower.*	Places 200 g on the round steel rod (50 cm long and 16 mm² cross section) and 200 g on the round steel rod (50 cm long and 10 mm² cross section). *You see, this one bends more because it is less wide.*
(d) *Bravo, can you demonstrate the same thing with others?*	*Yes* Uses square steel rod (50 cm long and 16 mm² cross section) instead of round steel, so the comparison is no longer exact. *This one bends more, it is less heavy.* Points to the narrow round rod.
(e) *And can you demonstrate the role of the form?*	Puts 200 g on the rectangular brass rod (50 cm long and 16 mm² cross section).
(f) Points to round, steel rod. *Why does this one bend more?*	*Because it is round.*
(g) *Is that the only reason?*	*The brass is also heavier.* Spontaneously discards the steel rod and takes a square brass rod 50 cm long and 16 mm² cross section.

FIG. 7.10 Protocol from the bending rods experiment (Kra, 14 years, 1 month). (Adapted from Inhelder & Piaget [translated by Parsons & Milgram] 1958. Copyright 1958 by Basic Books, Inc. Reprinted by permission.)

allowing only thickness to vary (c). But he is not quite sure of his method. He next (d, e) tries allowing material as well as form to vary, but he finally corrects himself and controls for material (e).

For Inhelder and Piaget, both Dur and Kra represent borderline cases between concrete and formal modes of thinking. They have a nascent understanding of the need and method for separating variables, but they do not apply it systematically. These children's responses are considered valuable in pointing out the gradual nature of the development of formal structures of thinking. Because they perform correctly once does not mean that they are well in command of the structures involved. It takes extended practice and experience for new logical structures to develop. Furthermore, these children's responses to the experimenter suggest that it is precisely during this transitional period that prompts and other forms of "instruction" have their greatest effect. This observation has special importance for education, and we return to its implications later in this chapter. First, however, we consider the general status of Piaget's stage and structure theories in light of recent research from an information-processing perspective.

CRITIQUES OF PIAGET

Perhaps inevitably, American psychologists—trained as experimentalists, initially uncomfortable with structuralist arguments and analyses, and predisposed to viewing the environment as a major determinant of human learning and development—have sought alternatives to Piaget's theory. Their critiques have been of three major types. One set of criticisms surrounds Piaget's stage theory of cognitive development, questioning whether the idea of discrete stages, based on the emergence of certain logical structures, can withstand the test of closer analysis. A related line of criticism concerns the psychological reality of the logical structures at the heart of Piaget's theories, especially the—to experimentalists' eyes—rather loose linkage of the data to his conclusions about children's competence. Finally, many critiques are concerned with the claim, implicit in Piaget's writings, that children are biologically programmed for optimal development in their normal social environment and that instruction can do little either to accelerate or to enhance the quality of children's logical functioning.

The Problem with Piagetian Stages

Most reports on Piaget's work, especially those intended to inform people concerned with teaching and instruction, place primary emphasis on the sequence of developmental stages: sensorimotor, preoperational, concrete operational, formal operational. These stages are often presented as if they were discrete time periods in children's lives and as if they set clear limits on the type of thinking that could be expected during each period. However, an extensive analysis of stage theory and related data by Flavell (1971) makes it clear that the stages are not all-or-none affairs; that is, children can behave as if they think reversibly and therefore operationally on one task, such as conservation of number, and still appear to be preoperational on some other closely related task, such as conservation of weight. Piaget acknowledges this phenomenon and has given it a name, *horizontal décalage*. But naming the phenomenon does not explain why it occurs, and the fact that it occurs represents a fairly serious challenge to a strict stage theory.

According to Piaget's stage theory, the acquisition of certain basic logical operations or structures such as reversibility allows children to recognize that spatial transformations (without adding or subtracting any quantity) leave quantities unchanged. But if the logical operations are present that allow a child to recognize conservation of number, why do they not also make it possible to recognize conservation of weight? Clearly other factors besides logical structures must be at work—factors such as the amount of experience a child has had with a particular quantity or material or the extent to which the critical features are easily visible or available for direct measurement. One can argue, for example, that conserving number or liquid quantity comes early because children have extensive experience—even before school begins—with counting and with pouring things from one container into another. One might also argue that in the young child's life there are more occasions requiring attention to a numerical or measured liquid quantity than occasions requiring attention to a weight. For example, when children share candy or drinks they do so on the basis of equal numbers or equal quantities. Rarely do they have to share on the basis of weight, nor are they often asked to compare weights with any accuracy.

None of these interpretations is necessarily contrary to Piaget's fundamental theory of structure in thinking, for Piaget would agree that the ability to deal logically with phonemena in the world is a result of the child's informal interactions with the normal environment. But these arguments do weaken the notion that logical structures, once attained, are so pervasive as to cause a distinct qualitative shift in the child's thinking. Instead of trying to pinpoint a stage of concrete operations, it seems more sensible to think of logical operations as one of many kinds of thinking abilities children acquire as they develop. This view suggests that a logical structure must be applied and "practiced" in a particular domain of knowledge before operational thinking in that domain becomes possible. For mathematics the implication is that experience with mathematical tasks, as opposed to more generalized experience with the environment, is more likely to improve children's ability to apply logical structures to their dealings with number, space, class inclusion, and the like.

The Difficulty of Assessing Competence

The notion that competence on specific tasks is a function of the growth of many skills is echoed in critiques that are not directly concerned with Piaget's stage theory. Some critics of Piaget raise the question: What does failure on a particular experimental task allow us to conclude about what children do and do not know? This question has often been formulated in terms of a distinction between *competence* and *performance*. (See, for example, the discussions in Brainerd, 1978.) Piaget's interest is in the child's underlying competence, that is, in the logical structures that the child commands. But we cannot directly observe these logical structures. Rather, they must be inferred on the basis of the child's performance on particular tasks. Does failure to perform a particular version of a problem reflect a lack of competence with respect to the presumed underlying logical structures? Or can the failure be attributed instead to the salience of certain misleading cues that do not clearly communicate to the child what question is being asked? An article by Trabasso, Isen, Dolecki, McLanahan, Riley, and Tucker (1978), reviewing evidence from their own studies and those of other investigators, illustrates this line of criticism. The review shows that variations in presentation of the class inclusion task can make large differences in children's likelihood of recognizing that a superset has more items than any one subset.

Trabasso et al. (1978) argued that a child might understand hierarchical relationships in general (e.g., know that apples and pears are subsets of the superset fruits) but fail to recognize the applicability of hierarchical relationships to a particular display in an experiment. This would produce failure on the standard class inclusion problem. The researchers devised a number of experiments attempting to influence the likelihood that children would encode (i.e., notice or recognize) the hierarchical relationships. Simply having children label the sets ("These are cats. These are dogs. These are all pets.") was not enough to improve performance, but other manipulations did have an effect. For example, if the items used as subsets were considered *typical* of the superset in question (horses are typical of the superset animals, but flies are not), children were more likely to succeed on class inclusion. It also helped if there were two different supersets present (for example, fruits, with subsets of apples and pears, and vehicles, with subsets of trucks and cars). It especially helped to ask questions that demanded comparisons across the supersets ("Are there more pears or more vehicles?") as well as within them ("Are there more pears or more fruits?"). When the two different supersets were present, children answered one question about as well as the other, and the pear-fruit question was more likely to be answered correctly than if the vehicles were not present.

Other experiments showed that part of children's difficulty with the standard Piagetian form of the class inclusion task might be in understanding the question. A study was described in which black cows and white cows were the subsets. In the standard condition the children were asked, "Are there more black cows or more cows?" It was hypothesized that the absence of an adjective for the superset might invite young children implicitly to insert one. Since the only available contrast adjective was "white," they would transform the question to "Are there more black cows or more white cows?" and make the standard class inclusion "error." If the experiment was run with the cows either lying down or standing up, however, the experimenter could ask, "Are there more black cows or *stand-*

ing cows?'' Adding this adjective greatly increased the proportion of correct responses. Other features, such as the size ratio of the subsets or the number of subsets shown for each superset, also influenced the probability of responding correctly.

All this variability argues convincingly that a child's failure on a standard Piagetian class inclusion test cannot automatically be taken to mean that the child has not developed a concept of class inclusion. Correct performance on a class inclusion task could depend on a number of separate variables—knowledge of which objects belong in which superset classes, the tendency to respond to linguistic constructions in adult-like ways (for example, without adding extra adjectives), the tendency to analyze displays hierarchically, etc. All of these factors must be taken into account, it is argued, to understand cognitive development, and the various abilities are not likely to develop simultaneously. Thus, the finding that a child has trouble comparing wooden beads with brown ones (the class inclusion task described earlier) should not be taken as evidence that the child has no concept of class inclusion. Instead the child may be missing any one of several different abilities or understandings that it takes to perform this particular version of the task. This does not mean, of course, that logical structures play no role in developing abilities or that they do not develop in a clear sequence. It does suggest that the status of certain logical structures that Piaget views as of primary importance in developmental growth may be less scientifically firm than many believe. And it certainly suggests that to design instruction based exclusively on Piagetian theory and analysis is to miss many important aspects of a good developmental theory of instruction.

Instruction on Piagetian Tasks

The interpretation of performance on class inclusion in terms of an interacting set of variables including but not limited to the use of certain logical structures is generally supported by a fairly extensive literature on attempts to teach children to perform the Piagetian tasks. This training research has had several motivations. Some American psychologists coming out of a behaviorist tradition were disturbed at the apparent implications of Piaget's theory for education. The theory seemed to imply that instruction could be effective only *after* structural changes associated with a given stage of development had occurred and, further, that these structural changes would emerge only as a result of very general kinds of experience—in other words, explicit instruction could make little difference. Some of the training research was aimed at disproving this aspect of Piaget's theory and showing that direct instruction could make a difference in tasks that were accepted as developmental landmarks. However, the psychologists who conducted most of the training studies were not interested primarily in the educational implications of Piaget's work. Rather, they were concerned with the mechanisms by which the environment brought about changes in cognitive com-

178 7. PIAGET AND THE DEVELOPMENT OF COGNITIVE STRUCTURES

petence and with the specific things children had to learn in order to become operational thinkers. To analyze the roots of operational thinking in terms of component skills and knowledge instead of focusing on the holistic characteristics of thinking seems a very un-Piagetian strategy. Indeed this was the reaction of Piaget and his colleagues to most of the early instructional experimentation. It was expressed quite pointedly in a lecture Piaget gave in New York in 1967, which was quoted by Elkind (1970):

> If we accept the fact that there are stages of development, another question arises which I call "the American question," and I am asked it every time I come here. If there are stages that children reach at given norms of ages can we accelerate the stages? Do we have to go through each one of these stages, or can't we speed it up a bit? ... It is probably possible to accelerate, but maximal acceleration is not desirable. There seems to be an optimal time. What this optimal time is will surely depend upon each individual and on the subject matter. We still need a great deal of research to know what the optimal time would be [p. 24].

Despite Piaget's disinterest, the American attempts at teaching Piagetian tasks yielded a great deal of information about how thinking abilities develop. They are also an interesting example of how psychologists working in one tradition can be enriched by contributions coming from quite a different tradition. In this case, the influence has gone both ways. Piaget has stimulated research on aspects of thinking that had not been considered before by American psychologists. At the same time, it is in large part because of American insistence on the importance of understanding the details of cognitive development that scholars from Geneva, working in Piaget's own research center, have turned to the problems of instruction (e.g., Sinclair, 1973). Let us consider some examples of these instructional experiments.

Gelman (1969) reasoned that in conservation of number tasks, and in other conservation tasks as well, the children acted as nonconservers (preoperational thinkers, in Piaget's terms) because they did not pay attention to the right features of the situation. In particular, they failed to recognize that there were several quantity dimensions, each of which had to be considered independently and yet simultaneously. Consider liquid quantity as an example. In a typical test for conservation of liquid quantity, the child is shown two glasses and asked to verify that they have the same amount of water in them. If the glasses are identical in shape and size, even very young children have no difficulty with this. But then the liquid from one of the glasses is poured into another glass that is, say, taller and thinner. Up to a certain point in their development, children typically say that the tall thin glass has more liquid, even though they have watched the pouring operation and know that nothing has been added or taken away. In this situation, there are several quantity attributes that the children could be attending to: overall size of the container, height of the liquid, width, etc. Yet

Critiques of Piaget   179

nonconservers appear to attend only to one of these, height. Could training be devised that would help children direct their attention to the most important of the attributes and to consider two or more of them simultaneously (i.e., separate the attributes and combine them appropriately)? This was Gelman's experimental problem.

The technique Gelman chose was to give children extended practice in making "same" and "different" judgments according to a named quantity dimension.

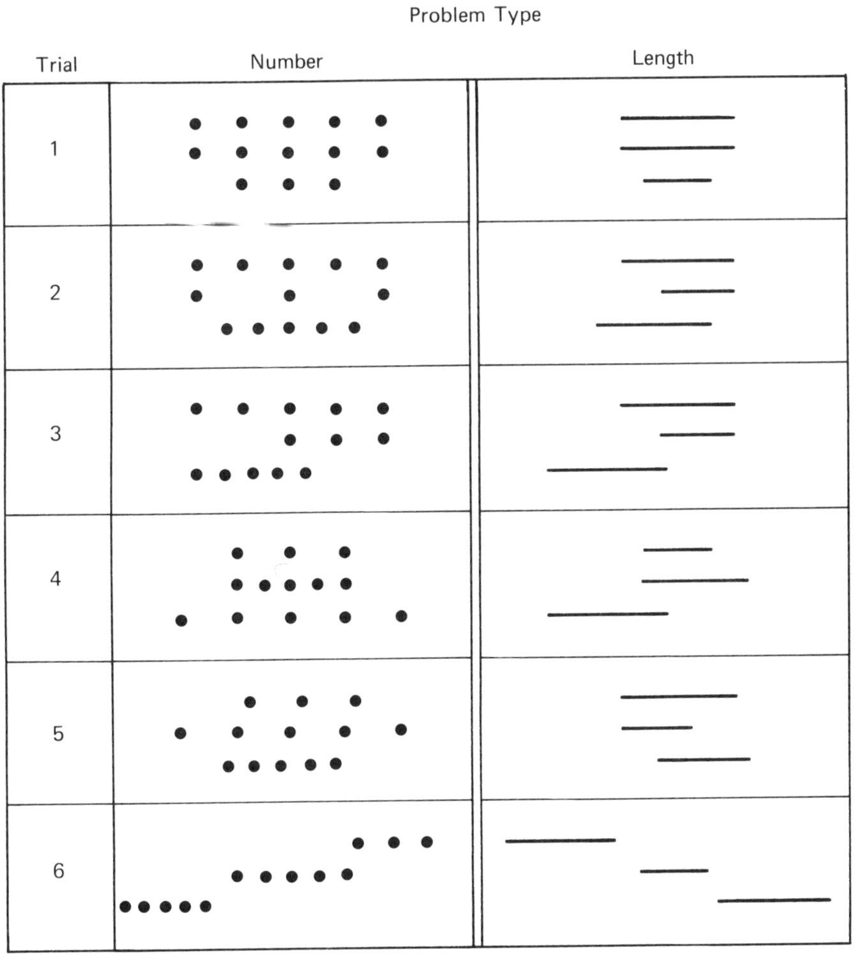

FIG. 7.11 Varied arrangements of stimuli (chips or sticks) used for practice in training conservation of number and length. (From Gelman, 1969. Copyright 1969 by the American Psychological Association. Reprinted by permission.)

Problems involving number and length were interspersed to show the children that a dimension that was relevant one time might not be relevant the next; that is, they must always consider the question being asked as well as the way the display looks. Figure 7.11 shows the kinds of stimulus arrangements that were used. For a number problem the experimenter would say, "Show me two rows that have the same number of things," or, "Show me two rows that have a different number of things." For length problems the task was, "Show me two sticks that are the same (or different) in length." After each response, children were told whether they were correct and were given trinkets as prizes for correct answers. After extended practice (about 100 trials each for number and length) virtually all the kindergarten children in the experiment gave the right responses 95% of the time. Thus, they had successfully learned to discriminate different quantity dimensions. Did this mean they would perform better on conservation tests than they had at the beginning of the experiment?

On the length test 18 of the 20 trained children scored perfectly, and on the number conservation test 19 scored perfectly. This contrasted with much lower scores for children who had not been told whether they were correct during training or who had been trained on a picture-matching task that did not involve quantity discriminations. Most of the trained children continued to perform perfectly 3 weeks later on a retention test. Gelman also tested the children on other conservation tasks—liquid quantity and mass—for which they had not been trained. Among the trained children, there was considerable transfer of the discrimination skill to these very different kinds of tasks. Although this transfer was far from perfect, it nevertheless indicated that children had acquired the habit of examining quantities carefully and that this could transfer to other dimensions as well. As in the class inclusion experiments, then, we have evidence that if children attend to or encode the relevant aspects of the stimuli, they can solve a problem that Piaget would have concluded was beyond their logical capacities. But this study, unlike the ones discussed by Trabasso et al. (1978), suggests that *some* fundamental principle was acquired, even if it was not the concrete logical operations claimed by Piaget. The tendency to conserve after training was carried over to dimensions other than number and length, so the children must have learned not only which dimensions to attend to but also the general concept that physical transformations are irrelevant to quantity.

A study by Bearison (1969) showed even wider transfer across different conservation tasks and demonstrated retention of the effect for 6 months—long enough for us to infer that it was probably a permanent concept for the children and not just a short-term effect such as is common in many training experiments. Bearison reasoned that if children could be made to understand the principle of measurement (i.e., that a single unit is applied repeatedly to a quantity), then they would be freed from the influence of perceptual features in the immediate displays when making quantity comparisons. They would understand that if two displays had the same number of units, then the quantities were equal regardless of how the units might be arranged or presented.

Bearison used the liquid quantity problem as the basis for his experiment. In the initial phase of training, white-colored liquid (called milk) was poured into many small 30-milliliter beakers—the kind that are sometimes used for dosing out medicines—and various numbers of these beakers were assigned to the experimenter and to the child. On each trial, the child had to decide who had the most milk on the basis of the number of filled beakers each person had. He was thus taught that quantity, even a "continuous" quantity like fluid, could be compared on the basis of a measured number. In the second phase of training, either the child's or the experimenter's liquid was transferred to a single larger beaker. The child did the pouring, counting the beakers as he poured in each one. Then, with the empty beakers still present so that they could be counted, the child had to compare the quantity of the experimenter's liquid with his own. There were numerous trials, in which the experimenter and child sometimes had the same number of beakers and sometimes an unequal number. Before passing on to the next phase, children had to demonstrate not only that they could decide correctly who had the most milk but also that they could justify the decision in terms of the number of separate small beakers involved. In the third phase of training, both the experimenter's and the child's beakers were transferred into larger containers, and comparisons were made as in the previous stage.

In the fourth phase the small beakers were not used; rather, milk was assigned to the experimenter and the child in two large, identical containers. Milk from one of the identical containers was then poured into another container of a different shape. It should be apparent that phase 4 amounts to Piaget's own conservation test, and quite a conceptual leap is required. In the experiment, children were permitted to go back and forth between the phase 3 training and the phase 4 conservation test as many times as needed to help them see the connection between the two situations.

Not surprisingly, the children who received Bearison's training learned to conserve liquid quantity, whereas untrained children did not. More interesting is the fact that the trained children learned to conserve in various other dimensions as well. The percentage of conservers was highest for "discontinuous" (discrete) quantity (same training procedure but using beads instead of liquid in the containers). Number, length, and mass (amount of clay) were also conserved by a significant number of the trained children, as compared with a control group. The 7-month posttest yielded the most persuasive data of all. There was some improvement in the number of conservers in all areas among both experimental (trained) and control children. But the trained children maintained their advantage in every area tested, including two—continuous and discontinuous *area*—in which they had not shown any advantage on the earlier test! Here we see strong evidence not only of retention but also of a kind of generalization that suggests a

7. PIAGET AND THE DEVELOPMENT OF COGNITIVE STRUCTURES

very fundamental principle was acquired and, over time, applied to more and more situations by more and more of the children. About three-quarters of the children were conservers 7 months after the training, even though they were only kindergarteners.

A Neo-Piagetian Analysis

We spent a good deal of time describing the Gelman and Bearison experiments because they are examples of successful instructional efforts in an area that for a long time had known few successes. The pages of developmental psychology journals are filled with reported attempts to teach children Piagetian tasks, but many of these studies failed or had very limited success: They showed only short-term effects and little or no transfer to related tasks. What do the successful training studies have in common, and what seems to account for their success? Many answers to this question have been offered, but few psychologists have attempted to analyze Piaget's theory specifically in terms of its implications for instruction. An exception is found in the work of Case (1978), who has proposed both a sequence of analytic steps and a set of variables with which to analyze Piagetian tasks.

Calling himself a "neo-Piagetian," Case explicitly accepts the fundamental features of Piaget's theory of cognitive development. He argues, however, that the difficulty of extrapolating from Piagetian theory to instruction arises because the theory is *structural* rather than *functional*. Piaget's protocols of children performing tasks offer us still snapshots of cognitive structures at various levels of functioning. They do not offer the kind of detailed psychological model of the acquisition and development of structures that is needed for designing instruction. In other words, Case calls for process models of Piagetian structures and of the transitions they undergo during specific learning episodes. In order to specify the nature of such transitions, one needs to analyze the mental steps required in cognitive tasks and to assess children's current performance on those tasks at various levels of development. This corresponds to the rational process analysis and empirical analysis that we describe in Chapters 3 and 4. Detailed work of this kind on Piagetian tasks has been carried out by information-processing psychologists: Klahr and Wallace (1972, 1973) have simulated conservation and class inclusion behavior on computers; Baylor and Gascon (1974) and Young (1973) have developed computer models that attempt to explain the behavior of children on seriation tasks, including their typical errors. None of these efforts, however, has directly addressed Case's concern, the analysis of Piagetian tasks for instructional purposes.

Case argues that the difficulty children have with particular tasks is a function of (1) mental actions or *schemes*[1] they have available to them; (2) the number of

---

[1]*Scheme* is a term used by Piaget to describe organized patterns of behavior (mental *and* physical) that occur in relation to the environment, a small-scale version of a structure. Case adopts the term in

schemes they are able to activate simultaneously; and (3) the strategy used in performing the task. All of these factors interact, of course, and the way they manifest themselves is a function of particular features of the task presentation. Other factors that influence task performance are the tendency of individuals to be influenced by the surrounding perceptual field ("field dependence" or "field independence"), the salience of relevant information in the problem situation, and the child's disposition or motivation to resolve the cognitive "conflict" (i.e., the discrepant information arising from an interpretation of the task features).

From a developmental point of view, the most important feature of Case's theory is the notion that as they grow older, children can activate an increasingly large number of schemes simultaneously. Building on prior work of Pascual-Leone (1970), Case defines a construct called the *M-space* (for *memory space*). This corresponds in a rough way to the concept of limited working-memory capacity discussed in Chapter 2, but the schemes that constitute "chunks" in M-space theory are larger than those in many working memory experiments. Different tasks require different numbers of schemes for successful solution, but complex tasks can sometimes be solved by using strategies that reduce the number of schemes needed simultaneously. Of course, discovering a simplifying strategy for oneself would require more M-space than just applying such a strategy. This means that if specific strategies are taught, they may make it possible for younger children (with smaller M-spaces) to perform tasks normally performed only by older people. Thus, specific knowledge affects one's performance level. This, Case argues, is why certain training studies work. Changing the task—for example, so that fewer competing schemes are activated by the visual display—can also help people with smaller M-spaces. This might account for why task modifications such as those described by Trabasso et al. (1978) for class inclusion could make such a difference.

Case's analysis of the liquid conservation task gives a sense of how these various factors interact. This is the task in which water from one of two equal beakers is poured into a tall, thin beaker and the child compares one of the original beakers (A) to the new, tall beaker (B). There are several ways of responding to this task. The following is one example:

1. Scan the vertical dimension of the water in Beaker A.
2. Scan the vertical dimension of the water in Beaker B (noting that the water in B continues past the point where the water in A ends).
3. Recognize that Beaker B contains a taller column of water.
4. Conclude that Beaker B contains more water than Beaker A [Case, 1978, p. 178].

---

describing mental behavior at the level of specific tasks. In connection with a liquid conversation task, for example, Case refers to a *figurative* scheme (keeping in mind how the two original beakers looked); various *operative* schemes (scanning the two beakers; storing information in long-term memory); and an *executive* scheme (a control function that keeps in mind the series of steps for solving a problem).

184   7. PIAGET AND THE DEVELOPMENT OF COGNITIVE STRUCTURES

This response is typical of very young children, who respond almost exclusively in terms of height. With respect to Case's theory the important point is that only a single scheme (the operation of comparing heights) must be attended to. If height could not be directly (i.e., visually) compared, more M-space would be needed, because two heights would have to be separately noted and then compared. Presumably, Case would also appeal to the M-space construct to account for the greater ease of number conservation when the sets are small enough to be subitized in parallel: Less M-space is required.

Another way of dealing with the liquid conservation task is to scan both dimensions, but without considering them simultaneously:

1. Scan the vertical dimension of the water in Beaker A.
2. Scan the vertical dimension of the water in Beaker B (noting that the water in B continues past the point where the water in A ends).
3. Recognize that Beaker B contains a taller column of water.
4. If the difference is large, conclude that B contains more water than A; otherwise, proceed to step 5.
5. Scan the horizontal dimension of the water in Beaker B.
6. Scan the horizontal dimension of the water in Beaker A.
7. Note that Beaker A contains a wider column of liquid.
8. If the difference is large, conclude that Beaker A contains more water than B. Otherwise, recycle to step 1, setting the criterion for "large" at a lower value [Case, 1978, pp. 179–180].

This strategy requires two simultaneous schemes, because the width comparison cannot be made on a direct visual basis. Further, if height and width differences were to be compared, three schemes would be needed, because height would be stored while the width comparison proceeded. A strategy of this kind would be required for noticing the "conflict" between height and width: Beaker A contains a wider column of liquid but Beaker B contains a taller column—they cannot *both* hold the larger quantity of liquid. In Piaget's and in Case's theories, such cognitive conflict is assumed to be the the incentive behind the mental work that leads to transitions to new levels of development; children work to resolve the conflict. In the ordinary environment, then, a relatively well-developed M-space that could accommodate up to three schemes (typical at about 7–9 years of age according to Pascual-Leone, 1970) would be needed before a child could spontaneously develop liquid conservation. Extrapolating to instruction, we might suggest training that would draw attention to the conflict by highlighting specific features of the stimuli (this is what Gelman's [1969] training did, for example) or by teaching specific strategies for making comparisons (as Bearison's [1969] did). But unless a strategy were taught that actually called on fewer schemes, Case's theory would predict no dramatic decrease in the age at which children could learn to perform specific tasks.

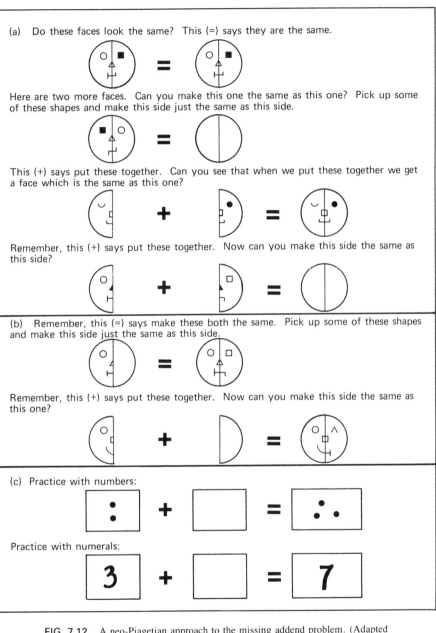

FIG. 7.12 A neo-Piagetian approach to the missing addend problem. (Adapted from Case, 1978. Copyright 1978 by Lawrence Erlbaum Associates, Inc. Reprinted by permission.)

Case has shown that tasks can sometimes be simplified dramatically by teaching the fundamental concept in a way that reduces the number of schemes that must be handled simultaneously and by teaching a relatively simple solution strategy. An example is the solution of missing addend problems, such as 4 + □ = 7, a task common in first-grade mathematics texts, but one that first-grade children find very difficult. Case worked with kindergarten children, presenting the series of exercises shown in Fig. 7.12. The first part of the procedure (a) familiarizes the student with the basic form of the missing addend problem. It uses only familiar elements (shapes, faces), and it maximizes the salience of cues (equation format, operational signs). It also reduces the number of schemes involved by eliminating the need to read numbers, to count, or to hold numbers in mind. The instructional emphasis is on matching both sides of the equation. The second part of the procedure (b) is designed to show the child that the common strategy of adding the first and last terms of the equation (4 + □ = 7; answer: 11) is counterproductive to the goal of matching both sides of the equation. Finally, the third part of the procedure (c) reintroduces the complexities originally stripped from the problem (such as numbers and counting). It provides reminders of the meaning and importance of the relevant cues (+ and = signs) and practice in executing the required task. Case also notes that young children who are successful in solving missing addend problems use the procedure of counting up from the given addend rather than subtracting it from the sum (cf. Groen & Poll, 1973). This requires fewer schemes than the adult strategy of subtracting 4 from 7.

PIAGET AND INSTRUCTION

It is no accident that our first discussion of instruction in this chapter came in the context of critiques and reinterpretations of Piaget's theory rather than in the presentation of the theory itself. In his own work, Piaget has not focused on questions of instruction. He considers himself a student of human knowledge and development, not of educational design. It falls to those of us who are concerned with instruction, therefore, to try to determine the implications of Piaget's theory for the teaching of mathematics. Many lessons have been drawn from Piaget and a surprising variety of specific educational practices and recommendations attributed to the influence of his theory. In the remainder of this chapter we consider some of these proposals, asking in what sense they are strictly Piagetian in origin, and how useful they seem as instructional guidelines.

Accelerating Progression Through the Piagetian Stages

Anyone who becomes familiar with the details of Piaget's research cannot help but be impressed by the ingenuity of the tasks used in his experiments. The tasks involve high levels of mathematical and logical thinking, in forms that appear capable of capturing and maintaining the interest of children. It might seem natural, therefore, to take the tasks as important in themselves and to set about developing curricula and teaching strategies that will accelerate the age at which children can perform them. To the extent that the tasks clearly reflect different stages of intellectual functioning, such a program would have the effect of accelerating progression through the stages of development described by Piagetian theory.

As we have seen, laboratory studies demonstrate the possibility of teaching various specific Piagetian tasks, with degrees of transfer and retention that suggest stable changes in cognitive functioning. Furthermore, whereas most studies show little effect of specific school experiences on acquiring more complex operational modes of thinking (e.g., Almy, 1970; Beilin, Kagan, & Rabinowitz, 1966), some studies in less industrialized countries suggest that appropriate forms of mathematics (Prince, 1968) or science (Duckworth, 1964) teaching can raise levels of performance on Piaget's tasks. Finally, there is evidence of broad cultural differences in the onset and timing of concrete operational thinking (see Laboratory of Comparative Human Cognition, 1979). Children from more industrially developed environments, with more specialization of labor and at least some schooling, tend to show evidence of operational thinking earlier than those from rural and unschooled areas.

All this suggests—quite in accord with Piaget's interactionist theory of intellectual development—that it might indeed be possible to design curricula and use teaching methods that would accelerate operational forms of thinking. Educational experiments exploring this possibility should greatly interest psychologists who study how human intelligence responds to environmental influence in the course of its development. But even if such experiments were highly successful, it is not clear that such acceleration would be a particularly desirable goal for education. There are also practical considerations, illustrated by the training studies we described. Whereas they proved it is possible to treat Piagetian tasks as subject matter—that is, to teach them as other mathematical content might be taught—they also demonstrated that doing so is likely to involve extended and complex forms of instruction. Consider that in both the Gelman and the Bearison experiments all instruction was done on a one-to-one basis, one child at a time. On a larger scale would such a costly effort be worthwhile?

The crucial question is whether, by accelerating development, we could improve people's chances of becoming, in the long run, better as well as earlier thinkers. Our current evidence suggests that people in all cultures sooner or later acquire the abilities necessary for at least concrete operational thinking. They all learn, without explicit instruction, to classify objects along more than one dimension, to conserve, to analyze geometrical figures in terms of interrelated structural features, and so forth. They do not have to be taught these things explicitly; they seem to acquire them through the normal transactions of living, even in quite "undeveloped" societies. Perhaps, then, we need to ask ourselves whether it is worth spending instructional time on things that will almost certainly be learned anyway. If the answer is no, as we believe it is, then focusing mathematics instruction on the acquisition of concrete operations would be a mistake.

For formal operations, however, the case is not so clear. Formal operations have been much less widely studied than concrete operations. If anything, there is even less clarity about which tasks in which forms actually reflect the structures that Piaget attributes to formal operational thinking. Nevertheless, it is clear that there do exist complex modes of reasoning, often associated with scientific thinking, that are difficult to attain and to use. Several investigators (Laboratory of Comparative Human Cognition, 1979) have concluded that a few cultures never arrive at fully developed formal operational thinking of these kinds. More important for our purposes, some individuals do not learn formal modes of thinking even in the most intellectually advanced cultures. Here indeed, if this evidence is correct, is a challenge worth considering. Can educational programs be developed that foster the development of formal logical modes of thinking, so that more people have an opportunity to achieve the levels of intellectual functioning now achieved by only a relative few? Can more people learn scientific modes of reasoning based on logic and mathematics? What would programs with such goals look like?

Although studies addressing this question are scarce, there is evidence that certain kinds of instruction may make it possible for preadolescent children to solve problems requiring formal operations. Siegler and his colleagues (Siegler, & Liebert, 1975; Siegler, Liebert, & Liebert, 1973) have taught children to sort out the effects of different variables on Inhelder and Piaget's (1958) pendulum problem. In this experiment, children observe the effects of varying weights and lengths of string upon the motion of a pendulum. The speed of the swinging motion is an effect of the length of the string to which the pendulum weight is attached, the actual weight being irrelevant. Inhelder and Piaget's 10- to 12-year-olds tended to attribute the speed to the weight of the object attached to the string. Siegler provided his subjects with conceptual training (definitions of dimensions and levels of factors, and the applicability of those concepts to problem solution) and gave them analogous problems to solve. Building a conceptual framework for scientific thinking and exercising the new concepts in analog tasks allowed these 10- and 11-year-olds to solve the pendulum problem quite readily. Case (1978) reports an experiment in which children as young as 8 years learned to separate variables in the bending rods task we described earlier. The method involved highlighting the separate variables (for example, by actually weighing rods and blocks) and providing a "stripped-down" version of the task for early practice. Results like Siegler's and Case's cast doubt on Inhelder and Piaget's implication that one must await children's readiness for formal operations training. At the moment, however, we have only a few experiments that hint at the possibilities for direct instruction in formal thinking. There is no organized curriculum to point to as an example whose effectiveness might be studied and evaluated.

Matching Instruction to Developmental Stages

There is another, more general way in which Piaget's sequence and stages of cognitive development might guide mathematics teaching. If his theory of development is correct, at least in broad outline, then it seems to set limits on the kind of reasoning and understanding we can expect from children at any particular point in their development. This implies that both the content and presentational techniques of teaching should be matched to the child's current level of development.

What would "matching" mean in instruction? How can the teacher respect children's current understanding and forms of thinking and at the same time play a significant role in their development? On first thought it appears that a curriculum designed to match children's developmental level should not expect them to engage in activities they are not yet fully able to understand. Thus, for example, one should not teach anything about addition until basic number concepts are well established and until laws such as commutativity and associativity are understood. Nor should one teach measurement until conservation of length is established. This might be called a "readiness" conception of how to match instruction to development. It assumes that fundamental understanding comes about through maturational processes or through some generalized environmental exposure.

Although this conception of matching has a distinguished history in psychology, it is virtually a counsel of despair for education, because it implies there is little for teachers to do except to await certain developments. If development is delayed or incomplete in some children, little is suggested in the way of educational activity to remedy the situation. Furthermore, the readiness conception gives little credence to the power of the environment, including the instructional environment, to influence the course of development.

A more positive definition of matching is available and has a growing basis in empirical experiment. It recognizes the ability, indeed the responsibility, of the teacher to influence important aspects of children's intellectual development. According to this conception, most fully developed by Hunt (1961, 1969), the important thing in education is always to pose problems that are slightly beyond the learner's current capability but not so far beyond that they are incomprehensible. Hunt draws on Piaget's notion of cognitive conflict, which is produced partly by the internal workings of intelligence and is often prompted by normal transactions with the social and physical environment. Cognitive conflict is what eventually presses individuals toward new and more powerful ways of thinking. He suggests that certain forms of instruction—defined as organizing the environment so that it makes new but possible demands—can foster structural reorganization and thus contribute both to the learning of specific new information and to overall cognitive development.

Lovell (1971) adopts a similar view:

> It is not in any sense suggested that the child must always be "ready" for a particular idea before the teacher introduces it. The job of the teacher is to use his professional skill and provide learning situations for the child which demand thinking skills just ahead of those which are available to him. It is a question of keeping the carrot just ahead of the donkey's nose. When a child is almost ready for an idea, the learning situation provided by the teacher may well "precipitate" the child's understanding of that idea [p. 17].

Lovell attempts to apply this principle directly to the problem of teaching mathematics during the primary school years. In keeping with his general understanding of Piaget's theory of intellectual development, the topics and tasks proposed by Lovell focus on the conceptual bases of mathematics and help children develop their understanding of these concepts; computational skill is not a central concern. His suggestions cover topics in number and set, operations on mathematical sentences, geometry and space, mappings, and pictorial representation. However, these topics and tasks are only loosely tied to Piaget's theory and research. Lovell acknowledges the difficulty of outlining Piagetian teaching implications for specific mathematical tasks: "Piaget's work cannot tell us which mathematical ideas should be introduced to children; only mathematicians and teachers can do that." The choice of topics and tasks, in other words, is defined by the subject matter of mathematics. But Lovell feels Piagetian research and theory can guide how they are taught by suggesting difficulties that children are likely to have at certain points in development and by providing a general framework of instructional principles.

For both Hunt and Lovell, then, the value of Piaget's work for matching instruction to learner capabilities lies not in any task he has devised or in the specific developmental stages he outlines. It lies in his general characterization of the quality of children's thinking and the instructional principles that can be derived from it. However, although the instructional principles proposed—such as constructive learning and the use of concrete representations for concepts—seem congruent with Piaget, they do not seem to derive uniquely from his theory. Indeed many of the "Piagetian" principles will be familiar from our description of structure-oriented teaching materials and methods in Chapter 5. It is nonetheless useful to review briefly several of these principles here, concentrating for the moment on their derivation from Piagetian theory.

Constructive Learning. "To understand," Piaget (1973) has said, "is to invent," to build for oneself. Although children can be helped to acquire mathematical concepts by means of special materials and teachers' questions, it is only through their own efforts that they will truly understand. Constructive learning thus implies activity by the learner, but activity of a special kind. The "active responding" called for in behavioral approaches to teaching is designed to provide an occasion for reward; the expected responses are largely specified by the teacher. In contrast, the activity called for by Piaget centers on trying to develop one's own approaches to particular tasks and problems. It is activity in which errors may be frequent, but these errors are part of the child's attempt to make sense of concepts. Constructive learning involves "trying out" ideas, testing to see which solutions work and which do not. This requires learning materials and learning environments that provide feedback to the individual on the outcomes of these trials. For Piaget, the kind of feedback that helps in this constructive learning process includes information from both the physical and the social environment.

Concrete Representations. As Piaget's experiments demonstrate, young children are able to think operationally only with respect to actually present materials and situations. They require feedback from the physical environment, in the form of concrete representations of concepts. And yet our educational system often depends almost exclusively on verbalization of ideas, in both teaching and testing. According to Piaget, verbalization does not ensure understanding, nor does understanding depend on verbalization. This being the case, instruction in a purely verbal mode is bound to fail, particularly when new concepts, demanding reorganization of thought structures, are being taught. In contrast, pictorial and concrete representations of mathematical concepts provide direct feedback on the correctness of children's tentative understandings. Thus, the various structure-based teaching materials cited in Chapter 5 may be taken as examples of Piagetian principles, although they do not directly derive from Piaget's work nor use the tasks characteristic of Piaget's research.

The Social Environment for Learning. The social environment provides the second kind of feedback that prompts children to give up old conceptions and structures and build new ones. According to Piaget, the kind of structural reorganization that is integral to the process of intellectual development comes about, in part, when children encounter disbelief of their proposals. As their social world expands with increasing age, they discover that other people do not always agree with their view of reality. Presumably this discovery leads children to examine more closely their own beliefs, to engage in closer observations and tests of the physical environment, and eventually to revise their conceptual structures. In this process, Piaget suggests, the disagreement of adults is less influential than the disagreement of children who are close to them in age and general conceptual level. If this is the case, then children's learning depends to an important degree on the social environment and the opportunity it provides to interact with peers over intellectual tasks.

To test this notion, Murray (1972) examined the effect of informal discussions among children upon the development of a basic concept, conservation. Murray pretested kindergarteners and first graders on six different conservation problems. He then organized the children into groups of three, with two conservers and one nonconserver in each group. Each group was required to arrive at collective answers to a series of conservation problems. There was no direct attempt to train the children in conservation; they were told only to discuss each problem, to explain to each other why they held the opinions they did, and eventually to agree upon an answer. A week later, every child was posttested on the conservation problems, this time alone. All the children improved their performance, the original nonconservers most of all. Because some of the posttest problems were quite different from the training problems, it is likely that at least some of this improvement could be attributed to social interaction and the idea testing induced by it.

The notion that social interaction plays a role in inducing the cognitive conflict that is the precursor of intellectual growth has often been cited in support of "open" or informal classroom environments in which extensive informal child-to-child interaction is possible. No good evaluations of the effects of such environments exist, however. In any case, their effectiveness would undoubtedly depend on the extent to which adequately structured and developmentally appropriate tasks are available for children to work on. Within the constraints of ordinary classrooms, providing structure without the straitjacket of simple drill exercises has so far proved an elusive, although still desirable, goal for educators in mathematics as in other fields.

Teaching as Clinical Interaction. Piaget's research is built on a special kind of interview technique, examples of which have appeared in the course of this chapter. The strategy is to set a distinct problem, embodied in physical objects that the child experiments with in the course of the interview. Both verbal responses and physical actions provide the data from which thought processes are inferred. The interview method itself, although not often thought of in this way, may constitute one of Piaget's greatest contributions to education. It provides a means by which teachers can understand what children understand. This is a crucial step in an educational strategy that seeks to match instruction to children's development. Teachers can cultivate their own skills of observing and questioning, as well as of setting interesting problems. As they become better at this, they begin to note details of children's thinking that had not been apparent before and find themselves able to follow children's lines of reasoning more clearly. Under these conditions, mistakes are not seen as poor thinking but as information about each child's current understanding. On this basis, tasks and questions can be posed that represent the best match in terms of the intellectual "stretching" propounded by Hunt (1961, 1969) and Lovell (1971).

This kind of clinical teaching could be applied in any substantive area, but it should not be thought of as independent of subject matter. Rather, it requires the kind of solid understanding of the subject matter that allows the teacher to recognize sensible but unusual responses and to invent problems that probe a child's understanding. Only teachers who themselves understand the conceptual bases of the mathematics they teach will be successful at the clinical technique. This understanding must come from studying mathematics, perhaps in ways specially adapted to questions of pedagogy. It will not come from studying Piaget or any other psychological theory that does not directly address mathematics learning and thinking.

SUMMARY

The developmental theory of Piaget emphasizes the dynamic aspect of intellectual activity and the psychological structures characteristic of children at different levels of development. In the writings of Piaget, the term *structure* is used as a means of describing the organization of experience by an active learner. Protocols of children engaged in mathematical and logical tasks are interpreted as evidence of qualitatively different cognitive structures that permit different understandings and solutions of the tasks. These differences in cognitive structures are said to develop in a sequence encompassing defined stages. During the normal period of schooling, children typically pass from the stage of preoperational thought through the stage of concrete operations to the stage of formal operations. Concrete operational thinking implies being able to reverse sequences of action mentally and to test various hypotheses. Formal operational thought implies an ability to think abstractly and to plan systematic variations in problem elements.

Critics of Piagetian theory question the reality of the stages, because there is so much variability in children's performances on tasks that presumably depend on the same operations. Noting the way in which task modifications can radically alter task difficulty, critics suggest that a number of variables, not only the logical structures that Piaget stresses, are required to account for performance; it is therefore not possible to infer a lack of logical competence on the basis of a particular performance. Neo-Piagetian theories have developed in an effort to define elusive structural concepts in terms of specific mental steps and strategies. This characterization makes Piagetian structures more amenable to precise definition and consequently to instructional intervention.

A number of attempts have been made to apply Piagetian theory to instruction. One approach is to teach Piagetian tasks as subject matter, in an effort to create the structures characteristic of operational thinking. Although successful training studies suggest this may be possible, the prospect of widespread teaching of this type is unlikely, given the high cost of extended training. The effort seems unwarranted for most children, because there is a high probability of their achieving at least concrete operational thinking through normal schooling and other experience. The development of formal operational thinking is less certain and therefore potentially more important for instruction. A few studies have shown that elementary school children can engage in formal operational thinking with appropriate training, but no test of this idea across a large array of tasks, as in a curriculum, has yet been undertaken.

A third instructional approach is to attempt to match instruction to children's developmental level. Rather than waiting for children to be "ready" for instruction, a more positive approach is to give children tasks that present something of an intellectual challenge but that have enough familiar elements so that they are comprehensible. General principles of constructive learning, concrete representations, social feedback, and clinical teacher–pupil interaction are derived from Piagetian theory. These can help to create optimal matches between learner capabilities and instructional content and procedures. We argue, however, that mathematics instruction is likely to benefit most from detailed psychological analysis of the content of the mathematics curriculum itself.

REFERENCES

Almy, M. *Logical thinking in second grade.* New York: Teachers College Press, 1970.

Baylor, G. W., & Gascon, J. An information processing theory of aspects of the development of weight seriation in children. *Cognitive Psychology,* 1974, *6,* 1–40.

Bearison, D. J. Role of measurement operations in the acquisition of conservation. *Developmental Psychology,* 1969, *1*(6), 653–660.

Beilin, H., Kagan, J., & Rabinowitz, R. Effects of verbal and perceptual training on water level representation. *Child Development,* 1966, *37,* 317–329.

Brainerd, C. J. *Piaget's theory of intelligence.* Englewood Cliffs, N. J.: Prentice-Hall, 1978.

Case, R. Piaget and beyond: Toward a developmentally based theory and technology of instruction. In R. Glaser (Ed.), *Advances in instructional psychology* (Vol. 1). Hillsdale, N. J.: Lawrence Erlbaum Associates, 1978.

Duckworth, E. Piaget rediscovered. *Journal of Research in Science Teaching,* 1964, *2,* 172–175.

Elkind, D. *Children and adolescents: Interpretive essays on Jean Piaget.* New York: Oxford University Press, 1970.

Flavell, J. H. Stage-related properties of cognitive development. *Cognitive Psychology,* 1971, *2*(4), 421–453.

Gelman, R. Conservation acquisition: A problem of learning to attend to relevant attributes. *Journal of Experimental Child Psychology,* 1969, *7,* 167–187.

Groen, G. J., & Poll, M. Subtraction and the solution of open sentence problems. *Journal of Experimental Child Psychology,* 1973, *16*(2), 292–302.

Hunt, J. M. *Intelligence and experience.* New York: Ronald Press, 1961.

Hunt, J. M. *The challenge of incompetence and poverty—Papers on the role of early education.* Urbana: University of Illinois Press, 1969.

Inhelder, B., & Piaget, J. *The growth of logical thinking from childhood to adolescence.* New York: Basic Books, 1958.

Klahr, D., & Wallace, J. G. Class inclusion processes. In S. Farnham-Diggory (Ed.), *Information processing in children.* New York: Academic Press, 1972.

Klahr, D., & Wallace, J. G. The role of quantification operators in the development of conservation of quantity. *Cognitive Psychology*, 1973, *4*, 301–327.

Laboratory of Comparative Human Cognition. What's cultural about cross-cultural cognitive psychology? *Annual Review of Psychology*, 1979.

Lovell, K. *The growth of understanding in mathematics: Kindergarten through grade three.* New York: Holt, Rinehart & Winston, 1971.

Murray, F. B. Acquisition of conservation through social interaction. *Developmental Psychology*, 1972, *6*, 1–6.

Pascual-Leone, J. A mathematical model for the transition rule in Piaget's developmental stages. *Acta Psychologica*, 1970, *63*, 301–345.

Piaget, J. *The child's conception of number.* New York: Norton, 1952. (Original French edition, 1941.)

Piaget, J. *On the nature and nurture of intelligence.* Address delivered at New York University, March, 1967. Quoted in Elkind, D. *Children and Adolescents: Interpretive essays on Jean Piaget:* New York: Oxford University Press, 1970.

Piaget, J. *Structuralism.* New York: Harper & Row, 1970.

Piaget, J. *To understand is to invent: The future of education.* New York: Viking, 1973.

Piaget, J., Inhelder, B., & Szeminska, A. *The child's conception of geometry.* New York: Basic Books, 1960. (Original French edition, 1948.)

Prince, J. R. The effect of Western education on science conceptualization in New Guinea. *British Journal of Educational Psychology*, 1968, *28*, 64–74.

Siegler, R. S., & Liebert, R. M. Acquisition of formal scientific reasoning by 10- and 13-year-olds: Designing a factorial experiment. *Developmental Psychology*, 1975, *11*, 401–402.

Siegler, R. S., Liebert, D. E., & Liebert, R. M. Inhelder and Piaget's pendulum problem: Teaching preadolescents to act as scientists. *Developmental Psychology*, 1973, *9*, 97–101.

Sinclair, H. Recent Piagetian research in learning studies. In M. Schwebel & J. Raph (Eds.), *Piaget in the classroom.* New York: Basic Books, 1973.

Trabasso, T., Isen, A. M., Dolecki, P., McLanahan, A. G., Riley, C. A., Tucker, T. How do children solve class-inclusion problems? In R. Siegler (Ed.), *Children's thinking: What develops?* Hillsdale, N. J.: Lawrence Erlbaum Associates, 1978.

Young, R. M. *Children's seriation behaviour: A production-system analysis.* C. I. P. Working Paper No. 245, Carnegie-Mellon University, Pittsburgh, 1973.

8 Information-Processing Analyses of Understanding

A recurring theme of the last several chapters has been that students of mathematics should be taught in ways that focus on an understanding of mathematical concepts—by making clear either the structure of the subject matter or the interrelations among elements of a stated problem. Bruner, Wertheimer, and Piaget would certainly agree that conceptually based representations of mathematical principles and problems facilitate mathematics performance. Studies of children's computational performances, too, pointed to the importance of conceptual understanding (see Chapter 4). Research on algebra word problems suggested that different problem representations had an impact on the success of solution efforts. Work on children's addition and subtraction algorithms clearly implied that even very simple arithmetic tasks, when efficiently performed, are rooted in a conceptual understanding of basic mathematical principles. Researchers showed that even errors are often signs of intelligent, although partial, understanding of basic principles. But none of the research we have discussed thus far has explored in detail the nature of conceptual representations. Nor has it explored the development of the understanding necessary to building and using them.

Can information-processing psychology, with its emphasis on sequentially ordered events, actions, and manipulation of information, shed any light on the question of how people understand mathematical concepts? Can it suggest how teaching might be organized so as to promote conceptual understanding? From the constructs and methods of modern cognitive research, can we draw insights into the relationship between conceptual understanding and performance of "routine" mathematical tasks, thus bridging the apparent gap between conceptual and computational approaches to mathematics instruction?

As we have seen, the question of how people understand is not new to psychology. It is, however, one of the newer topics within information-processing psychology and one of the most active and rapidly changing. This means that the theories we can point to today will almost certainly have undergone major revision in a decade. But it also means that the promise of this branch of psychology for bridging the boundary between computational and conceptual approaches to mathematics is only beginning to be realized. Not all cognitive psychologists agree on the promise of the constructs and methods we describe here—in fact, not even all who would call themselves "information-processing" psychologists do. But we think the promise is great. For the first time, psychology has a language and a body of experimental methods that is simultaneously addressing both the skills involved in performance and the nature of the comprehension underlying that performance. We attempt here to convey the flavor and possibilities of these new forms of psychological analysis.

In this chapter, we touch upon topics representing a few of the many directions information-processing psychology has taken in the past few years in the search for explanations of understanding and other such general intellectual capacities. Some of the methodologies are basically the same ones described in Chapter 4—reaction time studies, computer simulation, protocol analysis—but the content to which they are applied is more complex. Working at the leading edge of our knowledge, these efforts are pushing the limits of information-processing psychology, testing its power to explain human thinking. Their focus is on the structure of knowledge within the mind and on the mechanisms by which knowledge is manipulated, transformed, and generated in the process of solving the myriad problems humans face both in adapting to practical demands of their environments and in following their more intellectual pursuits.

First, we introduce the notion of semantic memory and network theories of how information is stored and organized. Like many other information-processing efforts, this work grew from attempts to program computers to simulate human capabilities—in this case, humans' ability to understand natural language. We then present an example of semantic analysis applied to the domain of mathematics and consider the role of such analyses in defining objectives for mathematics instruction. The problem of assessing the quality of knowledge is discussed, and several criteria are introduced for judging the degree of structure present in peoples' memory for specific topics. Next, we consider the role of knowledge and understanding in the context of problem solving, including the influence of problem representation, the task environment, and strategy upon the application of knowledge. Finally, we describe a series of instructionally oriented experiments to illustrate the use of knowledge and strategy in specific problem-solving situations.

THE ORGANIZATION OF SEMANTIC MEMORY

To deal with understanding in information-processing terms, we need to enlarge and elaborate our conception of the nature of long-term memory. Recall that within most information-processing models a distinction is made between *working memory*—where coded information is temporarily stored for immediate use and where active processing of the information goes on—and *long-term memory*, or what we sometimes call *semantic memory*. Long-term or semantic memory is where everything the individual knows is stored—permanently, according to many theories.

But how is it stored? And how might the way it is stored affect our ideas about how to foster understanding? A simple possibility is to view the contents of long-term memory as a long list. In that case, to retrieve information one would mentally "scan" the list, rejecting item after item until the needed information was found. However, a little reflection on how many things most people know makes it clear that simply searching through unordered lists would be an exhausting and inefficient way of retrieving information. The size of the list would be enormous, and people do not search fast enough to make this a reasonable model of human performance. Further, a simple list-search model does not account for how people make inferences or, indeed, how they deal with any but externally presented ideas or relationships. For example, if an item were "forgotten" or became confused with some other item on the list, there is no way, in a pure list-search model, for the individual ever to "figure out" what the lost item was or to reconstruct new information.

For information-processing psychology to deal with the capacities of humans to understand, to generalize, and to invent, a richer conception of how people store and retrieve knowledge has had to be developed. To account for the complex range of people's ability to recall information, to make inferences, and generally to use their knowledge, it seems clear that we must conceive of our memories—our stores of knowledge—as organized and structured. And so, although there are a number of different theories or models of semantic memory (e.g., Anderson, 1976; Anderson & Bower, 1973; Norman & Rumelhart, 1975), all of them describe human knowledge as structured and organized. In this sense they resemble the gestalt model of the human mind. In other ways, however, current information-processing models seem more similar to associationist conceptions. Let us attempt to illustrate this dual character of structuredness and associativity with some simple examples.

When semantic memory theorists talk about how information is organized, they tend to speak in terms of specific "knowledge structures." Figure 8.1 shows a piece of a hypothesized knowledge structure concerning animals, following a model developed by Collins and Quillian (1969, 1972). The structure is composed of units that are nouns—in the present case, the name of a general category "animal" and the names of some specific animals. Each unit has

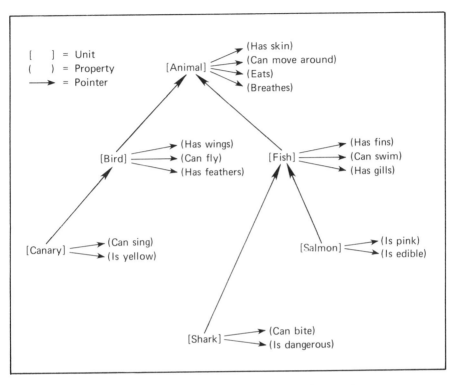

FIG. 8.1 Portion of a hypothesized knowledge structure in semantic memory, showing relationships among the units and properties within the category "animal." (From Collins & Quillian, 1969. Copyright 1969 by Academic Press. Reprinted by permission.)

associated with it certain properties. Properties refer to physical features of the unit (e.g., "is yellow," "has skin"), things the unit can do ("can bite," "eats"), things that can be done to it ("is edible"), and more general characteristics ("is dangerous"). In this model, these units and properties are ordered hierarchically. However, whether they are really hierarchically ordered in the human mind is debated in the literature, and most semantic memory models specify networks of associations that are not necessarily arranged in strict hierarchies. The relations between items in the network are shown by pointers. Thus, *Bird,* and *Fish* both point to *Animal,* whereas *Shark* and *Salmon* point to *Fish.* Links between items in the network are associations, not so different from Thorndike's "bonds." But in this model—as indeed in all current semantic memory models—the association is based on a particular relationship (X *has* [object]; X *does* [action]; X *is* [adjective]). The number of possible relationships is finite, and this in itself lends a coherence to knowledge that simple contiguity associations, bonded together only by frequent occurrence together, would not have. Structure is also specified in the particular network relationships, which place certain items of knowledge in central, or "nodal," positions and organize knowledge in interrelated chunks.

A semantic network theory of knowledge can account for a number of mental capabilities. It can, for example, account for both the kinds of statements people are willing to make about any particular topic (in the present case, about animals) and the relative speed with which they make them, as measured in reaction time experiments. For example, although there is no direct pointer relating *Canary* to *has skin,* people are quite willing to agree that canaries have skin. But it takes them a fraction of a second longer to do this than it does to agree that a canary has wings. According to the theory, this is because one has to go through more pointers, starting at *Canary,* to reach *has skin* than to reach *has wings.* Thus, the greater the "distance" between concepts, the longer it takes to decide whether a relationship exists between those concepts.

Semantic structure theories can also account for how people make inferences. For example, most people have probably never thought before about a canary's skin (presumably because they are more likely to notice the feathers than the skin on a bird). But they can figure out that, because the canary is an animal and animals have skin, canaries do indeed have skin as well as feathers. Thus, network structure models account for how people know things without having to list every separate item of knowledge. They reflect the assumption that the human mind can *construct* knowledge—which we can view as finding new relationships among concepts—as well as receive knowledge from events in the environment.

According to semantic memory theories, then, the human mind is active, not a passive recorder of associations from outside. Somehow in the course of development, knowledge becomes structured in meaningful ways; it is not just a random collection of bits of information. A certain similarity of this view to those of Bruner, Piaget, and the gestaltists is no accident, for the psychologists who are developing semantic memory models are seeking to explain some of the same characteristics of human thinking that these earlier theorists noted. They are attempting to do so, however, in terms of rigorously stated models that have certain characteristics: (1) They are specific to the content of the knowledge domain under study. (2) They can incorporate both rules for action and conceptual relationships within the same networks. Because of these characteristics, semantic memory models permit the theoretical linking of computational procedures and knowledge of the principles underlying these procedures. (See Brown & Burton, 1978, and Greeno, 1978b, for discussions of "procedural networks.") For the moment, at least, it seems possible that much of what we call *structure* in mathematical thinking can be accounted for in the neo-associationist terms of semantic networks. At the very least, it is worth vigorously pursuing this relatively new stream of psychological thinking, even as we remain alert to other possible ways of characterizing human knowledge that may extend the range of our explanatory power and perhaps eventually our instructional power.

A Knowledge Structure for Arithmetic Operations

Let us consider an example from mathematics to see what the semantic memory notions suggest about what kinds of knowledge will be easiest to learn, to use, and to remember. Although originally linguistic and psycholinguistic rather than mathematical in content, the semantic memory work has recently been taken up in the domain of mathematics and logic. It seems likely that mathematics, because of its relatively more structured and transparent content, will be one of the first domains in which psychology will truly be able to make contact with instruction. Our examples of mathematical knowledge structures are adapted from Greeno's (1978b) consideration of the relationships between multiplication and division.

Figure 8.2 shows a pair of possible knowledge structures for multiplication and division. Note that our representation includes not only units and properties but also explicitly named relationships. Items (nodes) in memory are related in specific ways, so that it is not simply a matter of ideas being associated or bonded together but of their holding a particular relationship to each other. The pointers between nodes thus bear descriptive names of their own. The directions of the pointers indicate the direction of relationship, such as subject to object, but imply no direction of processing; it is assumed the links can be traveled in either direction during memory work.

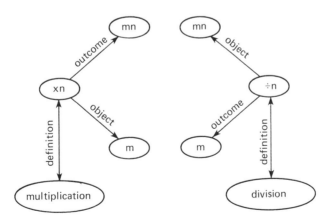

FIG. 8.2 Possible knowledge structures concerning multiplication and division. Their separation indicates nonunderstanding of the inverse relationship between the two operations.

According to the structures shown in the figure, multiplication is defined as the operation "times-n" ($\times n$). The operation $\times n$ has an *object* (the thing upon which the operation is performed), in this case the quantity (m), and an *outcome*, the quantity (mn). Division is also defined as an operation, $\div n$; and this operation too has an object and an outcome. The figure represents the knowledge structure of someone who knows about multiplication and knows about division but does not understand their inverse relationship. The multiplication and division structures remain unjoined.

To understand that multiplication and division are the inverse of each other means recognizing that there is a special relationship between the object quantities and the outcome quantities for the two operations. Specifically, if a person multiplies a quantity by some number (say n) and then divides the outcome by the same number (n), the original quantity is again obtained. This understanding is shown in Fig. 8.3, where the outcome quantity for one operation is represented as the object quantity for the other. The knowledge structures of multiplication and division are now joined and the total structure is simplified in the process.

As we have already suggested, this view of knowledge builds structure from the connections among ideas. Recall that, although based upon the associationist notion of bonds as the contents of the mind, the present theories do not treat all bonds alike. Each connection implies a particular relationship, and these relationships in turn imply different ways of using the connection. A person thinks differently, for example, about a connection that says one thing *is equivalent to* another (as in the relationship between multiplication and $\times n$) and a relationship that says one thing is the *object* of another (as in the relationship between $\times n$ and m). Similarly, the *object* relationship is different from the *outcome* relationship. Thus, items in memory are not just *connected* to each other; they are

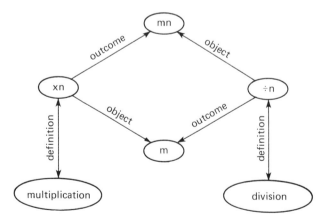

FIG. 8.3 A knowledge structure showing understanding of the inverse relationship between multiplication and division. This represents a joining and simplification of the structures in Fig. 8.2.

related to each other in definable and therefore meaningful ways. Another aspect of this conception of knowledge that is illustrated here is its recognition of multiple connections. Every item in the knowledge structure is a node in a large network, with (potentially) several pointers leading into and out of it. Each pointer represents one of the many kinds of things one can know about that item.

Let us expand the representation of multiplication and division to show this multiple connectedness a little more fully. Figure 8.4 shows how a person with a well-developed conceptual structure for arithmetic might understand the four operations of multiplication, division, addition, and subtraction. Addition and subtraction are defined as operations, each having an object quantity and an outcome quantity. Like the relation between multiplication and division, the inverse relation between addition and subtraction is understood because the same quantities—a, m, and $m + a$—play object and outcome roles in both operations. The diagram also shows that the addition and subtraction operators are related. Specifically, the operation "times-n" is further defined as equivalent to successive addition (plus-m done n times) and "plus-m" is represented as a component of this defining operation. A similar definition of division as successive subtraction is also shown.

Another kind of understanding is represented in this knowledge structure. We do not analyze the operation here; but notice that addition is shown as the incrementing operation we described in Chapter 4 in connection with reaction time experiments on the addition process. This is an example of the way experimental data can contribute to a theoretical semantic analysis. In the addition experiments, it may be recalled, the performance of 5- and 6-year-old children indicated that they felt quite free to use either m or a as the starting quantity. For this reason, addition is defined as either of two operations, plus-m or plus-a, differing in their object quantities but identical in their outcome quantities. This is one possible representation of a person's knowledge of commutativity in addition. Careful inspection of the figure shows that it also represents knowledge of the *non*commutativity of subtraction. On the subtraction side of the diagram, minus-m and minus-a start with the same object quantities but produce different outcome quantities. The diagram shows a similar structural base for understanding commutativity in multiplication and noncommutativity in division.

It is easy to see that one could expand the figure to include considerably more knowledge about arithmetic operations and their interrelations. For example the figure could be made to show the relations between addition and subtraction that produce the decrementing-down and incrementing-up algorithms for subtraction that we see in second- and fourth-grade children in Chapter 4. And it could show alternative definitions of addition, subtraction, multiplication, and division—for example, addition defined as movements on a number line rather than incrementing operations or multiplication defined in terms of rectangular arrays. As our hypothetical individual's conceptual knowledge of arithmetic grew, so would the number of linking relationships in our structural representation. But only a few

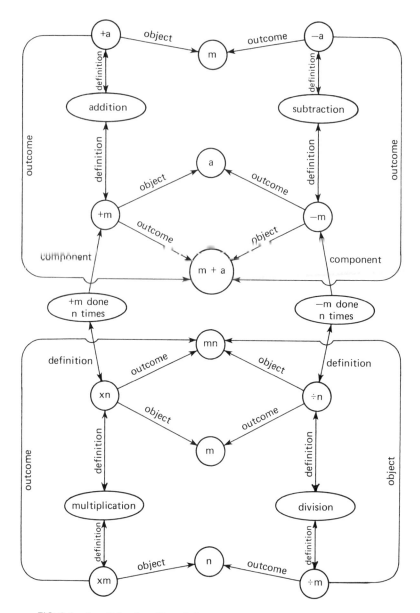

FIG. 8.4 A well-developed knowledge structure for the four operations of addition, subtraction, multiplication, and division.

204

more nodes might be added, because learning "more" usually results in better organized and linked knowledge rather than more separate pieces of information. Thus, the person whose knowledge could be represented by the structure in Fig. 8.3 knows more about arithmetic than the person represented in Fig. 8.2. Yet the structure in Fig. 8.3 is actually simpler—because greater knowledge has allowed the same nodes or items of information to function in both the division and the multiplication knowledge systems.

Well-Structured Knowledge as a Goal of Instruction

We are leading, of course, to the notion that a major goal of mathematics instruction ought to be helping students to acquire "well-structured" knowledge about mathematics. Later we discuss the relation of knowledge structures to performance of mathematical tasks and the contribution of well-structured knowledge to insightful problem solving in mathematics. But first we need to establish some criteria for well-structured knowledge to guide our thinking.

According to Greeno (1976), the real usefulness of semantic network representations for instruction lies in their specificity. They allow us to specify objectives of instruction at a level that takes into account what people *know* in relation to what they *do*. These are what Greeno has called "cognitive objectives," as opposed to purely behavioral objectives. In outlining cognitive objectives, we do not discard performance or behavior as criteria of educational attainment. Rather, we extend the notion of formal educational objectives to include the knowledge structure that produces the behavior as well as the behavior itself.

What this extension entails can be seen most directly by contrasting semantic network diagrams with learning hierarchies, described in Chapter 3, and even with the more detailed process analyses of mathematical tasks and performances in Chapters 3 and 4. Learning hierarchies result from analyzing tasks into target behaviors with both stimuli and responses specified (e.g., *given* a fixed, ordered set of objects, *the child can* count the objects). Complex acts are broken down into the component skills that are included in them, but hierarchies do not specify exactly how the various skills, procedures, or ideas are interrelated in a person's mind such that skilled performance results. Process analyses, both rational and empirical, describe what might be called "procedural knowledge"—people's internal programs for performing mathematical routines—but they do not specify the conceptual knowledge of mathematics to which these procedures are linked. Cognitive objectives, particularly when specified in terms of semantic networks, can suggest the kinds of knowledge structures, both conceptual and procedural, upon which instruction should focus. They can provide a theory of knowledge to guide instruction, a theory suggesting that there is more to mathematical performance than the performance itself and that the same overt performance may be based on different understandings of the concepts involved.

206   8. INFORMATION PROCESSING ANALYSES OF UNDERSTANDING

The task of specifying cognitive objectives will be a major undertaking for research in the psychology of mathematics learning. To do the job well and usefully it will be necessary to combine conceptions of the subject-matter structure, general standards of well-structured knowledge, and empirically derived information about the knowledge structures of people who are recognized as expert in a domain. Generally speaking, we are much better at describing ideal knowledge structures—those based on rational analysis of the subject matter, such as the structures for arithmetic operations described earlier in this chapter—than we are at verifying whether these knowledge structures are actually the ones that guide the performance of real people. Since we cannot look into people's minds and observe network diagrams, knowledge structures must be inferred from their behaviors. Many standard research methods—measurement of response latencies, study of error patterns, and analysis of thinking-aloud and performance protocols—now play a role in inferring people's knowledge structures. Measures of association among words that make up subject-matter terminology are also being employed. Further, techniques of clinical interviewing, similar in intent and style to those of Piaget but analyzed in difficult terms, are being adopted as information-processing psychologists increasingly turn their attention to knowledge structures. Later we describe empirical studies using these various methods, but first we pause to consider the problem of outlining general criteria for well-structured knowledge, criteria that can be used to guide and inform empirical work as it proceeds.

Greeno (1978b) has suggested three criteria that can be applied in evaluating the degree of understanding reflected in a semantic system. These are (1) internal integration[1] of the representation; (2) degree of connectedness of the information to other things the person knows about; and (3) correspondence of the representation with the material that is to be understood. Let us try to understand these criteria in the context of the diagrams for knowledge about the four arithmetic operations (Fig. 8.2-8.4). *Correspondence* refers to a good match of one's mental picture with those held by experts in the field. In this case it means having mathematically "correct" definitions of each of the operations, "correctness" being defined by the consensus of mathematicians. Figure 8.4 shows a structure that corresponds to that of mathematicians—for example, in its clear depiction of the commutative and noncommutative aspects of the four operations—although expert mathematicians would of course know even more about the operations than the figure represents. The contrast between Fig. 8.2 and 8.3 illustrates an improvement in internal *integration,* because the parts that were initially separate, the object and outcome quantities, are now seen as related to each other, representing the inverse relation between multiplication and division. The figures do not directly show the quality of *connectedness.* To do so, they would have to

---

[1]Greeno termed this internal "coherence," but because he did not imply *logical* coherence, we thought it better to use a term with no such connotations.

show the variety of problem situations in which concepts and operations of arithmetic were called upon, how the problem solutions were related to the operational definitions, and how this knowledge structure was related to general mathematical and logical concepts. A knowledge structure meeting the criterion of connectedness would allow a person to answer interpretive questions and apply the structure in new situations because of its links to other knowledge.

It would be nice if we could point to well-developed and tested ways of assessing these three criteria or, indeed, the characteristics of people's knowledge structures in general. An experimenter examining the relationship between knowledge and performance or an instructor trying to design teaching materials for a given mathematical topic both need to know what individuals' knowledge structures for those topics are—how coherent they are, how well connected to other related topics, and how much in correspondence to the structure of mathematical knowledge specified within the discipline of mathematics. But strategies for assessing knowledge structures are still rather ad hoc, designed for particular experiments and not well suited to formal or quantitative comparisons of instructional conditions and their effects. In a sense, the criteria of integration, correspondence, and connectedness constitute an agenda for the development of assessment procedures that take into account the knowledge structures of students. Meanwhile, some possible directions for such efforts can be suggested, partly by considering psychological studies of a related subject-matter domain, physics.

Assessing Integration. Consider first the criterion of internal integration. One measure of well-structured knowledge is the extent to which concepts are associated with each other in rich yet orderly ways. In an integrated knowledge structure certain concepts are central, in the sense that they are associated with an especially large number of other concepts. These are nodal or organizing concepts. According to semantic network theory, access time is related to the number of connections that have to be traversed to reach a particular concept or to find the relationship between two or more concepts. Under this assumption, a person whose knowledge is integrated, in the sense of having a few organizing concepts that link to a significant subset of other concepts, ought to show a certain pattern of behavior in timed free-recall experiments. When asked to talk about a topic or a problem (e.g., "Tell me everything you know about Newton's law of moments"), that person should show a pattern in which a number of items are retrieved in a "burst," followed by a brief delay and then a burst of further items. Presumably, the items within the burst are the ones grouped together in memory around a common organizing concept. The time between bursts is spent searching for the next organizing concept, which may involve traveling a long or complicated route of associations.

The general notion that internal integration affects temporal patterns for memory retrieval has been pursued by Larkin (1977) in studies comparing the information structures and problem-solving routines of novices (college students) and experts (their instructors) in physics. Larkin calls the information grouped around an organizing concept a "chunk," following terminology previously used in comparing the information-processing strategies of novice and expert chess (Chase & Simon, 1973) and Go (Reitman, 1976) players. According to Larkin, if information were being drawn from an unchunked (i.e., not very integrated) system such as we would expect a novice to have, the time elapsing between any two freely recalled items would be distributed randomly. For the expert, by contrast, many pairs of items would be separated by very short time intervals, with relatively fewer pairs separated by longer time intervals. The longer intervals would come at the points where new nodal concepts or chunks were being accessed; the shorter intervals would occur within chunks. Larkin compared the time patterns of a student and a physics teacher in recalling physics concepts and found exactly what she predicted. The expert appeared to be retrieving information in chunks. The novice on the other hand appeared to retrieve information more or less randomly.

Presumably, we might expect that over time spent studying physics, the novices—at least those who were learning well—would begin to look more like experts with respect to their chunking patterns. Thus one way to assess a student's progress in learning might be to examine temporal patterns of recall, looking for evidence of chunked information organization and, by extension, integrated knowledge. However, Larkin's method allows one to conclude only that information is chunked. It reveals nothing about which items are chunked together. Students might show more and more chunked recall patterns but still not match the content patterns of experts' knowledge. This is, in fact, what one might predict of students who "cram" for an exam. Such students may generate for themselves various mnemonics for linking items in a course syllabus together, but they do not necessarily discover the links that characterize the thinking of experts.

Assessing Correspondence. Measures of integration, then, are only useful if we also have a way of assessing the correspondence of learners' knowledge structures with the structure of the subject matter they are being taught. Again, a possible approach is to compare novices' structures with those of experts, but with attention to *which concepts* are grouped and which relations are expressed. Shavelson (1974; Shavelson & Stanton, 1975) and Geslin (1973, 1974), have developed word association, graph building, and card sorting tests as methods for determining the match between individuals' cognitive structures and the subject-matter structure as expressed by instructors and textbooks.

In the association method, a person is given a word drawn from the subject-matter domain and is asked to state as many other words or concepts associated with the target word as possible. Once a list of associated words is collected for each target word, it is possible to compare the lists for degree of overlap. Any pair of target words can be considered and a ratio constructed that compares the number of words common to both lists to the total number of common associates that are possible for those lists (this ratio is termed a "relatedness coefficient"). Relatedness coefficients for all possible pairs in a set of target words can be arranged in matrix form, and the matrix then can be analyzed statistically for clusters of concepts that go together. This clustering should make intuitive sense to those familiar with the subject matter. It can also be compared with clusters derived from a formal analysis of conceptual relationships in a text or other written document on the topic.

One would expect that, in the course of instruction, students' patterns of concept clusters would become progressively more similar to those in a text or to the clusters shown by an instructor. Shavelson (1972) tested physics students at six points in the course of their instruction and found that their cluster patterns moved closer and closer to the structure displayed in the instructional material. Students in the physics course also gave more associations to the target word than nonphysics students, and their associations were more central to the formal definitions of the physics concepts. Thro (1978) carried out a similar study in which both students and the instructor took the word association tests several times during a semester's physics course. Like Shavelson, she compared the matrices of relatedness coefficients of students and their instructor at each point and found an increase over time in the match between them.

An important additional finding in Thro's study concerned the relationship between cognitive structure, as measured by the word association technique, and performance on physics problems that required applying known equations in specific situations. Thro found that students who performed best on the physics problems had cognitive structures most closely matching the instructor's by the end of the course. This suggests that the degree of correspondence between a knowledge structure and the formal structures of the subject matter is directly linked to performance on semialgorithmic problems. This is what we would expect if performance routines are based on conceptual understanding of a subject matter or if understanding develops, in part, as a function of practice in using the procedures associated with a subject matter. However, this relationship between word association measures and ability to solve problems cannot be viewed as firmly established. In an earlier attempt to follow the development of cognitive structures, Rothkopf and Thurner (1970) compared essays on concepts from Newtonian mechanics written before and after reading a physics text. Analysis of the essays, using a method that assessed patterns of co-occurrence of key terms, showed that essays written after reading the text were somewhat more similar to the text than those written before. But there was no corresponding change in performance on a physics problem-solving test. Thus, people's knowledge structures—at least as assessed by verbal association measures—may begin to approximate those of experts without also yielding the new understanding that expands the learner's power to perform in subject-matter tasks.

Assessing Connectedness. It appears that we cannot depend entirely on association measures in studying cognitive structures. We also need to develop ways of inferring knowledge structures directly from problem-solving performances themselves and from interviews in which people are asked to explain the basis for their problem-solving activity. These methods, difficult to quantify but often clear in the interpretation they allow, may be of particular importance in helping to link procedural with conceptual knowledge and thus will constitute a way of studying connectedness as well as integration and correspondence of people's knowledge structures.

An example of this approach to studying knowledge structures can be given through the description of a child participating in a current study by Resnick. The study concerns children's learning of place value concepts and procedures. Leslie is a 9-year-old who systematically used a common faulty subtraction rule, namely, the smaller number is subtracted from the larger in each column regardless of its position in the subtrahend or the minuend. She used this routine both when subtraction problems were presented in standard, vertical calculation form and when they were embedded in story problems. In addition problems, on the other hand, she used a correct carrying rule. Thus it appeared she knew something about place value, but this knowledge seemed unconnected to her knowledge about procedures for subtraction. The lack of connectedness and integration in Leslie's knowledge of subtraction was highlighted by her performance on the following task, in which she was shown a pair of problems side by side:

$$\begin{array}{rr} 36 & 27 \\ -27 & -36 \end{array}$$

She was asked which problem would be easier to subtract and why and was then asked to do whichever one she wanted. Leslie claimed the two were identical and neither was any easier to do than the other. However, when asked to *do* the calculation, she *always* chose the problem with the larger number on top and then used her standard "buggy" algorithm. (For an introduction to the study of buggy procedures, see Chapter 4.) Asked why she chose the problem she did, Leslie said that her teacher always insisted on doing subtraction with the larger number on top, so she (Leslie) did it that way. Her oral responses clearly indicated she knew a rule about putting the larger number on top but did not have this rule connected either to procedural knowledge about how to perform subtraction calculations or to conceptual knowledge about place value and the logic of subtraction.

In the course of the study, Leslie was taught an appropriate borrowing routine for subtraction. Instruction began by using Dienes blocks (described in Chapter 5) in base 10. This allowed representations such as those in Fig. 8.5 and permitted exchanges (10 smaller for 1 larger or vice versa) to model carrying and borrowing. Leslie quickly learned the rules for representing numbers, interpreting block displays as numbers, and making exchanges. This seemed to indicate she already knew the fundamental meaning of place value, because she could

Training Procedure	Dienes Blocks Representation	
	Tens	Units
For the problem 85 − 47 1. Represent the 85 with blocks.		

2. Start in the ones column and try to remove the 7 blocks shown in the subtrahend.
3. There aren't enough blocks there, so go to the tens column and "borrow" a ten-bar.
4. On the written problem, cross out the 8, and write 7, to show the change in blocks in the tens column:
     7̸8̷5
   − 47		

5. Trade the ten-bar for 10 ones-cubes and place them in the ones column.
6. On the written problem, represent this by writing a 1 that changes the 5 to 15:
     7̸8̷¹5
   − 47		

7. Now remove the number of blocks shown in the ones column of the subtrahend.
8. Count the number of ones blocks remaining, and write the answer in the ones column of the written problem:
     7̸8̷¹5
   − 47
        8		

9. Go to the tens column and try to remove the number of blocks shown in the subtrahend.
10. Since there are enough blocks, complete the operation, count the blocks remaining and write the answer in the tens column of the written problem:
     7̸8̷¹5
   − 47
      38		

FIG. 8.5 Sample of the procedure used to train Leslie in subtraction using Dienes blocks.

translate block representations into standard written and spoken numerals, and vice versa, unbothered by zeros—a frequent stumbling block for children uncertain of place-value notation. The subtraction procedure shown in Fig. 8.5 was then demonstrated to Leslie. The demonstration was accompanied by a thorough explanation of why each step was taken. Following the demonstration, Leslie was led through the steps of several such problems, gradually doing more and more of the work herself. During this process, more complex problems were introduced, including three-digit problems. Eventually Leslie carried out the entire procedure herself and within 40 minutes was executing it smoothly. She even solved problems with zeros in the minuend with no difficulty. The Dienes blocks were then "faded out," through a process in which the experimenter pretended to manipulate imaginary blocks, as directed by Leslie. Within 10 minutes Leslie was solving problems in written form only.

To explore the extent to which Leslie had understood and incorporated into her own knowledge structure the routines she had been taught, she was brought back 3 weeks later for a follow-up test. On this occasion Leslie performed a series of written problems with ease. There were no mistakes; everything was done just as she had been taught. Thus, she clearly remembered the algorithm. To probe Leslie's understanding of the algorithm, she was asked a number of questions. Her responses suggested that she understood why the algorithm worked, especially as her explanations used wording different from what the experimenter had used during instruction.

Even stronger evidence for understanding, and for a now well-connected knowledge structure, came from Leslie's ability to subtract using a way of notating place value—expanded notation—that had not been explicitly tied to the subtraction algorithm. Here is a problem Leslie solved after a brief introduction to expanded notation. It is the first problem she encountered in which there were zeros in both the units and tens columns:

```
  700
 -356
```

L: Writes: 700 + 00 + 0
            300 + 50 + 6

"I have to go here (hundreds column) 'cause there is nothing to borrow from here (tens column)."

E: "You are borrowing a hundred from here (the 700). Where are you going to move it to?"

L: "Here (points to the zero in the ones column)."

E: "If you move it here, this (the zero) becomes a hundred. Then 6 from 100 would have to be what?"

L: "Ninety-four."

E: "How will you write that?"

L: Writes the 4 in the ones column.
Changes the 00 in the tens column to 90. Thus:

$$\begin{array}{r} 600 \quad 9 \\ \cancel{700} + \cancel{00} +_{10}0 \\ \underline{300 + 50 + 6} \\ 4 \end{array}$$

Leslie got into potential trouble by jumping to the ones column after borrowing. But she averted it by invoking a *carrying* rule that put the extra digit into the tens column. She thus combined carrying and borrowing in the same problem, and yet finished without difficulty, getting the answer of 300 + 40 + 4, or 344. The experimenter then showed Leslie a different routine (all borrowing, no carrying) for solving the expanded notation problem and asked Leslie how it could be that both methods yielded the same answer. Leslie accounted for the equivalent results in terms of a place-value interpretation of the numbers (e.g., "the 9 in 94 is really 90 and so belongs in the tens column").

Leslie's entire performance with expanded notation seems extraordinary for a child who only 3 weeks before had performed subtraction as if she had no understanding at all of place-value rules. The ease with which she learned the blocks routine for borrowing, her ability to transfer with only minimal help to a potentially confusing notation of place value, and her ability to combine borrowing and carrying within the form of notation all suggest Leslie knew something about place value before the experiment. But until she received special instruction, that knowledge had not been adequately connected with subtraction. We conjecture that the instruction linked the written subtraction algorithm to place-value knowledge Leslie already possessed. Once the connection was established, Leslie was able, largely on her own, to connect a different representation of place value with subtraction and thereby to invent a sensible, if unorthodox, subtraction routine.

USING KNOWLEDGE AND STRATEGY FOR SOLVING PROBLEMS

The unintegrated and unconnected character of Leslie's knowledge prior to instruction, and her subsequent ability to use her knowledge to solve a problem not previously encountered, point to the need for a psychological theory of understanding that goes beyond the presence or absence of specific knowledge elements. Stored subject-matter knowledge alone cannot solve problems. There must also be a mechanism to direct the mental search through the networks to retrieve information. And there has to be a mechanism for actively generating and testing new relations among concepts and structures when the needed information is not stored in exactly the form that seems to be required.

Information-processing theories conceive of the mind as possessing, in addition to knowledge structures, a repertoire of problem-solving *strategies* that help to interpret problems, locate stored knowledge and procedures, and generate new relations among separately stored memory items. These strategies organize the thinking process and call upon various components of knowledge to put together a plan of action capable of solving the task at hand. To account for problem solving in mathematics, then, we need to consider both the kinds of mathematical knowledge structures people have, including the kinds of algorithmic routines they are capable of performing, and the strategies they have for accessing their knowledge, detecting relationships, and choosing among the actions available to them. We discuss here three aspects of problem-solving strategy: (1) how problems are represented; (2) how features of the task environment interact with an individual's knowledge; and (3) how problems are analyzed and knowledge structures are searched to bring initially unrelated information to bear on a task.

Representing the Problem

The first step in any problem solving situation is to build a representation of the problem; that is, to notice features of the problem and encode them in such a way that they are interpretable by the information-processing system. The information given in the problem statement must be encoded in forms that match elements in the individual's knowledge structure. This is what allows one's knowledge to be applied to the problem at hand. Much of the work on representation in psychology has been driven by attempts to construct theories of how people understand language, how they attach meaning to the language that is heard or read by connecting words and sentences to already established knowledge structures. However, a few studies on the role of representations have been conducted in the domain of mathematics.

In our earlier discussion of algebra problem solving (Chapter 4), we noted evidence that constructing physical representations of verbal problems could allow a person to bypass complicated equation setting and solving. This was particularly evident in the case of physically "impossible" problems, but there is good reason to suppose that the effects of initial problem representation are much more general. Bruner (Chapter 5) hypothesized three different modes of representation—enactive, iconic, and symbolic—that seemed developmentally related in children but that might be drawn upon as the situation required in problem solving. Gestalt demonstrations (Chapter 6) pointed up the importance of developing problem representations that are rich enough to make contact with relevant knowledge. In the present context, we view building a representation as the process of establishing links between the problem statement, on the one hand, and the person's semantic network, known procedures, and general knowledge concerning mathematical and spatial relationships, on the other.

Let us examine the related processes of representation and solution in the context of a simple word problem:

> Sharon is 6 inches taller than Danny, who is 1 inch shorter than Jesse. If Jesse is 4 feet tall, how tall is Sharon?

Using Knowledge and Strategy for Solving Problems   215

How would a child go about solving such a problem? Presumably not by simply beginning to add or count or subtract. The child who understands how to deal with word problems would not begin operating on the given numbers without first figuring out what the *words* demand be done. Interpreting this problem requires a variety of linguistic skills—recognizing nouns, adjectives, and verbs and using various referential cues in the language. For example, the child must recognize that the word "who" in the first sentence refers to Danny and that "taller" and "shorter" are related in a special way. The linguistic processing serves to encode the problem for processing, relating the words of the problem to the person's own stored linguistic categories. In a sense, this means reformulating the problem "in one's own words." Beyond this general linguistic processing, a strategy is needed for identifying what is known and what must be found out.

Exactly how this is done depends on a combination of the individual's knowledge and preferred strategy. A child who does not know any algebra might solve the problem through a nonformal representation in which natural language (the words of the problem or the child's own translation of them) plays an important part. The following is a nonalgebraic solution sequence that can be characterized as informal linguistic processing. The material in brackets suggests some of the processing that is probably required to reach the statement:

> Jesse is 48 inches tall. [4 feet equals 4 times 12 inches. This formulation involves using stored knowledge about the relationship between feet and inches. It employs the strategy of translating given information into common units.]
>
> Danny is 1 inch shorter than Jesse. [The word "who" in the story refers to Danny.]
>
> So Danny is 47 inches tall. [If Danny is 1 inch shorter than Jesse, then *subtract* 1 from Jesse's height.]
>
> Danny is 6 inches shorter than Sharon. [Inversion of the sentence, "Sharon is 6 inches *taller* than Danny."]
>
> Danny is 47 inches tall. [Recalled from prior processing.]
>
> So Sharon is 53 inches tall. [If Sharon is 6 inches taller than Danny, then *add* 6 to Danny's height.]

It is evident that this process involves considerable knowledge both about the general structure of the English language and about the specific ways in which quantitative and comparative information is expressed linguistically. This alone would suggest why children find word problems so much more difficult than corresponding calculations stated in simple numerical form. The difficulty is even more comprehensible when we consider how much the child who solves in this way must keep in mind as solution proceeds. In light of the limited capacity of working memory, it is not surprising that children sometimes forget steps or forget the outcomes of calculations already performed, even if they understand perfectly well the language of the problem. Other solution strategies that use some form of intermediate, nonlinguistic representation can help reduce this memory load.

For example, a physical representation of the problem might help in this respect. A mental image can be built of three children standing side by side, having approximate heights based upon the preliminary linguistic analysis of the problem. By visualizing the differences between their heights and assigning mathematical expressions to those differences, as in Fig. 8.6., the problem is readily solved. The visual representation holds the problem information in a (presumably) readily accessible form while calculations are carried out, thus reducing memory load and thereby the likelihood of errors.

Someone who knows algebra can use a formal algebraic representation as a way of reducing the memory load. This is accomplished by constructing the equations on the basis of linguistic information in the story but leaving all calculation to the end. Here is a possible sequence that translates the problem into algebraic sentences:

Sharon is 6 more than Danny:	$S = D + 6$.
Danny is 1 less than Jesse:	$D = J - 1$.
Jesse equals 48 inches:	$J = 48$
Sharon equals what:	$S = ?$

The problem can now be solved by the rules of algebraic substitution.

There are, then, at least three possible representations of this simple problem—informal linguistic, physical/visual, and algebraic. Each representation has the potential for tapping different kinds of knowledge and calls upon

FIG. 8.6 Visual (iconic) representation of a word problem.

different strategies for solving the problem. The informal linguistic analysis simply applies operators to the numbers as they are suggested by the various restatements of the word problem; the visual representation calls upon knowledge of spatial relations and sets up the computation as a direct visual comparison; an algebraic representation calls up the rules for manipulating symbolic expressions in an equation. We are not suggesting that people represent problems in only one of these ways; indeed it is more likely that several ways will be considered and one or some combination of ways chosen for the final solution process. Further, some representations will be more successful than others for each given problem. The point is that different representations have the power to call up different facts and procedures from long-term memory, and this in turn will affect the solution process and the likelihood of its success.

What determines the kind of problem representation that is optimal or that is likely to be established for a given problem? The available possibilities depend, of course, on individual differences in the extent and structure of prior learning, including the developmental limitations imposed by age and experience. The nature of these differences was the subject of the first half of this chapter, where we discussed the value of well-structured knowledge that is internally integrated, well connected to other knowledge, and in good correspondence with agreed-upon principles of mathematics. A second factor influencing representation is the task environment, including task instructions, and the cues it contains that tend to call up one or another type of representation.

Task Environment

The task environment comprises all the elements of a task that are available to and perceived by the problem solver (i.e., the "givens" of a problem). The problem may be presented physically, as when a teacher hands Joey two sets of blocks and asks him to tell which set has more. It may be presented diagrammatically, as when an achievement test poses questions based on a graph. It may be presented verbally, as in word problems, or symbolically, as in solving for unknowns in an algebraic expression. More often the problem is presented as some combination of objects, pictures, symbols, and verbal instructions. The problem statement, whatever form it takes, provides the raw materials out of which the information-processing system builds a representation of the problem. This in turn determines which solution strategy is selected.

The givens of a problem in a sense demand that the problem solver do certain things; that is, they cue specific procedures and call up specific information. A pair of scissors, for example, cues the act of cutting; a square root sign cues the procedure of finding the square root of a number. These cues naturally lead to a problem representation that includes that kind of information or those procedures. But the problem materials also cue more global strategies that in turn affect the process of building a problem representation. For example, task instructions usually set up an explicit goal for problem solving (e.g., "prove that triangles *abc* and *def* are congruent" or "find the set of whole numbers that satisfy the equation, $3x + x - 10 = 0$"). The goal then directs the course of problem solution.

We have seen numerous other examples of the effect of the task environment upon global solution strategies in our discussion of gestalt principles. The principle that "the whole is more than the sum of its parts" implies that individual problem features are perceived in an organized way that imposes structure upon the problem as a whole. Sometimes the various elements of the problem give rise to erroneous structures and thus to false solutions. To take an example from Wertheimer (1945/1959, p. 29), a student starts in on an exercise page and shows the following work:

$$12 = 3 \times \underline{4}$$
$$56 = 7 \times \underline{8}$$
$$45 = 6 \times \underline{?}$$

To the last problem, which is a stuck question, the student confidently answers 7. Why? She has been misled by the problem arrangement. If you ask her for her reasoning, she responds, "Isn't it clear? The fourth digit is one higher than the third—12 3 4, 56 7 8, and 45 6 7". Apparently the task environment has told her something about certain number sequences but nothing about the intended demands of the problem. Perhaps the child has not yet come across the signs for "equals" and "times" in her arithmetic studies and thus is unable to encode those aspects of the task environment in a meaningful way. In any case, for her the task environment gives rise to a goal of "find the next number in the sequence" and not the correct goal of "find the number that multiplied by 6 yields 45."

The task environment is a powerful determinant of the range of strategies that a problem solver can bring to bear. This is amply demonstrated in experiments on "functional fixedness," a line of research growing out of the gestalt work on insight in problem solving. In these experiments, objects that have conventional uses must be used in novel ways to solve a problem. In Maier's (1930) two-string problem, for instance, two strings were hung from the ceiling in such a way that a person holding the end of one string could not reach the second string without dropping the first. The instructions were to tie the two strings together. In the room were placed several objects: poles, clamps, pliers, extension cords, tables, and chairs. Because of their conventional uses, the objects suggested certain applications to the problem. One way experimental subjects could solve the two-string problem was by tying the pliers to one string and swinging the string like a pendulum so as to bring it within range of the second string. But this solution depended on being able to view the pliers as a weight rather than as something to grip with. Many of the subjects were blind to this use of the pliers until the experimenter provided a subtle hint: "accidentally" brushing against a string while walking across the room, so that it swung back and forth suggesting a pendulum. People's tendency to think of objects as having fixed uses restricts the use of those objects in novel situations. By the same token, stimuli in the task environment can cue habitual solution strategies that may or may not be successful.

Task Instructions. Task instructions can be particularly effective as either aids or deterrents to problem solving because of their power to generate representations. An example is found in a recent experiment by Magone (1977) based on Maier's two-string problem. Three classes of solutions to this problem are possible, according to Magone's analysis: *extension* (of the string by tying another long object to it or of the arm by using some rigid long object to hook the string in); *anchoring* (holding one string down in the middle, while walking over to reach the other one); and *pendulum* (putting a weight on one string and swinging it toward the other). Some of Magone's subjects were required to use the anchor solution on the first try; other groups were told to use the extension and pendulum solutions first. They then had to solve the problem five more times, using each of the objects available, but with no restraints on the type of solution. Subjects in these three groups tended to continue using the kind of solution that they had used initially—for as long as there were objects available. They only switched to another type of solution when they had run out of objects, usable in the original solution type. Subjects in another group, who had not had the problem represented for them in a particular way, showed no particular patterning of their solution attempts. They simply used each object in the most immediately obvious way. Apparently, the differences in task instructions to these various groups brought about differences in problem representations—and in the tenacity with which they clung to one or another representation.

Some of the experiments designed to test the relative effectiveness of discovery learning as opposed to rule learning or to rote learning (see Chapter 6) can be interpreted as failures of task instructions to establish the kinds of representations that would lead to problem solution. Recall Katona's (1940) task requiring subjects to memorize a lengthy number series. Small wonder that the group that was asked to recite the numbers over and over compared unfavorably with the group that was asked to look over the number series until it was "well learned." The task instructions established totally different learning sets, or mental representations—if instructions were followed, only the "discovery" group had the opportunity to notice features of the task that led to recognizing the numbers as an ordered mathematical series. Similarly, the study by Gagné and Brown (1961) comparing discovery, guided-discovery, and rule-learning methods gave very different task instructions to the discovery and the rule groups, thus probably inducing very different representations of the problem. The following instruction contained in the guided-discovery program suggests why this group might have done so well:

> What do you have to do to obtain the sum, or Σ? Can you get Σ^n by adding, subtracting, dividing, or multiplying the numbers in the row pointed to? For exam-

ple, when Σ^n is 3 [in the series 1, 2, 3, 4, 5, 6, . . .] the term-value pointed to is 4. If you know 4, you can get 3 by _____ _____ [Gagné & Brown, 1961, p. 316].

Instructions such as these seemed to have the effect of forcing students to reinstate concepts such as:

term number, n, term value, T, T_{n+1}, which they had previously learned in the Introductory program, as well as some concepts learned even earlier, such as subtract, divide, multiply, and add. That is to say, the items specifically required [the subject] to respond by writing, or otherwise reinstating, these verbal responses. In contrast, it may be noted that the [rule and example] program did not require this; instead, entities like term number, T, T_{n+1}, occurred as stimuli (in the formula provided) to which the required responses were the locating and copying of specific numerals [Gagné & Brown, 1961, p. 319].

We can interpret this in terms of the process of building a problem representation. Instructions that facilitate building a rich representation by cuing previously learned concepts have a different effect than instructions that cue finding and copying examples of rules. Thus, the verbal instructions that accompanied the guided-discovery program were undoubtedly partly responsible for better transfer of learning to new situations.

The task environment, then, consists of the objective elements of the task, including the task instructions or problem statement, and physical elements of the task such as drawings, diagrams, or concrete objects. It should be clear from this discussion that the objective elements of any problem become transformed in the process of encoding the elements for mental processing. They become part of what Newell and Simon (1972) call the internal *problem space* or what we have been calling the *problem representation*. The problem space is actually a richer and more complete task environment than the external one, because it includes not only the givens of the problem, reformulated in familiar terms, but also all the relevant facts, relations, and procedures that these reformulated elements are connected to in long-term memory.

STRATEGIES FOR ANALYZING PROBLEMS AND SEARCHING KNOWLEDGE STRUCTURES

Once an initial problem representation is built up, the probability of correct solution depends on whether the solver has in memory an appropriate set of procedures that fit the problem as represented. With most problems involving algorithmic solutions (i.e., the typical problems presented in standard mathematics courses) the important intellectual work is over once a representation has been developed. One need simply ascertain that cued procedures are indeed applicable to the problem and then apply those procedures. Usually standard algorithms—

for example, those for solving algebraic equations—can be applied. But sometimes a problem, at least as currently represented, is not susceptible to a direct algorithmic solution. A problem solver may recognize, for example, that a known procedure is useless in the specific situation or that there is a gap in his or her knowledge. In such a case, the "problem" may be that the match between givens and stored knowledge is not obvious, or that the needed information is not available, or that there is a huge amount of stored information to be sorted through in order to locate the match. Often false solutions are generated because the student simply fails to recognize that a previously successful routine cannot work in the new problem situation. Once the inapplicability of known procedures is detected, however, a number of strategies are available to help locate stored information or to piece together the needed information from previously unassociated bits of knowledge.

Research on high school students' solutions of geometry proofs (Greeno, 1978a) illustrates the range of possible interactions between strategies and specific conceptual knowledge. Greeno collected thinking-aloud protocols from students as they worked on problems similar to those in standard high school geometry courses. On the basis of these protocols, he developed a computer program, PERDIX, that attacked geometry problems in much the way the students did; that is, PERDIX's solution responses closely matched the students' responses. Examination of the inner workings of PERDIX, therefore, provided strong clues to the students' thinking about the problems and to the kinds of information they were using.

A sample problem is given in Fig. 8.7. The problem, stated in panel (a), is to find the measure of $\angle q$, given that lines a and b are parallel, lines m and n are parallel, and the measure of $\angle p$ is 40 degrees. In panel (b), all the angles in the drawing are uniquely labeled (A1 through A16) for ease of reference. Panel (c) represents the knowledge state after the initial encoding or representation of the problem. The material in solid lines depicts what is known: at this point, the given measure of $\angle$ A1. The material in dotted lines depicts the information being sought: the measure of A12, which is to be expressed as a yet unknown number (?NUM).

PERDIX begins by making a series of tests to see if there is a direct solution. First, it determines whether any angles of known measure are congruent to A12. This test is possible because the system has knowledge that vertical angles and corresponding angles are congruent. It thus finds an angle vertical to A12, A15, and tests whether its measure is known. Since it is not, other angles vertical or corresponding to A12 are located and similarly tested. All of these tests fail in the present case, but PERDIX can try another fairly direct solution: If there is a supplementary angle (e.g., A16 or A11) whose measure is known, then A12 can be found by simply subtracting the known value from 180 degrees. To make this test, A16 is found and tested, but its measure is not known; neither is the measure of A11.

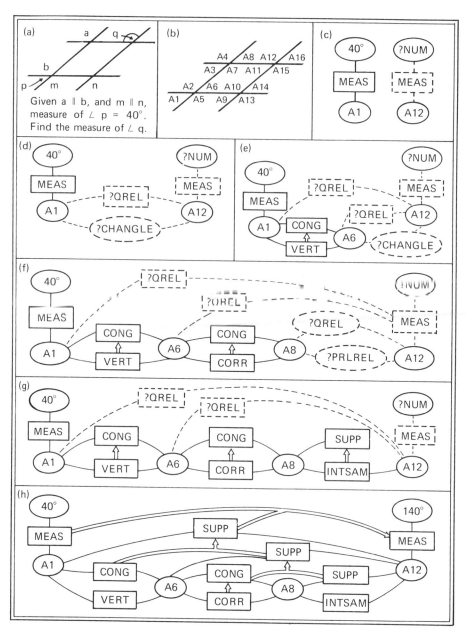

FIG. 8.7 Stages in the solution of the vertical angles problem. (From G. Greeno, 1978a. Copyright 1978 by Lawrence Erlbaum Associates, Inc. Reprinted by permission.)

This initial search for information that could yield the measure of A12 illustrates a strategy that is quite common in problem solving, often called the *generate-and-test* strategy. If one knows what kind of information is needed but does not know which items fit the criteria, it is possible to scan a list of candidate items, testing each one in turn to see if it fits. A very simple example of the generate-and-test method is seen in the problem of finding all the prime numbers between 1 and 50. Assuming a list of prime numbers is not already stored in memory, a person may proceed by generating all the numbers between 1 and 50, one at a time, and testing each one to see whether it meets the criteria for prime numbers (i.e., it has no factors other than 1 and itself). The numbers that do have other factors are eliminated. The remaining numbers constitute a list of prime numbers and hence a solution to the problem. The search can be shortened considerably by generating only odd numbers from 1 to 50, because all even numbers have a factor of 2. In other words, information about the structure of the numbers serves as a *heuristic*, a strategy for reducing the search space. In the PERDIX geometry problem, a heuristic is applied as well. Only angles with a known relationship to angle A12 are generated (first vertical angles and then supplementary angles). In this case, the search space is reduced by applying knowledge of the rules of congruence that are brought to mind in interpreting the visual pattern represented in the diagram. Variants of the generate-and-test strategy appear throughout PERDIX.

Continuing with the explanation of Fig. 8.7, because no congruent or supplementary angle with known measure can be found, PERDIX clearly needs a more complicated search plan. What PERDIX does is to set up a *subgoal*, namely, to find some quantitative relation between the target angle, A12, and the given angle, A1. Under this subgoal, a new series of tests is made. First, PERDIX determines whether a quantitative relation between these angles has already been inferred. Finding none, it tests to see whether there is a known positional relationship between the two angles, generating relationships such as vertical angles, corresponding angles, supplementary angles. A1 and A12 have no such relationship (all the tests are negative), so PERDIX sets up another set of subgoals. It begins to look for an angle forming part of a *chain* of angles that are quantitatively interrelated and that can thus link angles A1 and A12. The new subgoals are shown in dotted lines in panel (d). The subgoal ?QREL means "Find the quantitative relation linking A1 and A12"; ?CHANGLE means "Find an angle in the chain linking A1 and A12." PERDIX has a "scanning route" for problems of this kind that establishes a characteristic order for examining the angles; the particular scanning route that humans use would of course vary with individuals. As shown in panel (e), A6 is found to stand in the relationship of vertical angle to A1, and this permits the *inference*, shown by a double arrow, that the two angles are also congruent.

Now a new subgoal—find the angle that will make a chain linking A6 and A12—is established. Note, however, that the original goal of finding the relationship between A1 and A12 remains, because it is not yet satisfied. Continuing with the scan of angles, A8 is accepted as related to A6 by correspondence. This allows the inference, shown in panel (f), that A6 and A8 are congruent and sets up a new subgoal of linking A8 to A12 through a relationship based on parallel sides (?PRLREL). The relation of interior angles on the same side (INTSAM) and the inferred relation of supplementary angles (SUPP) are shown in panel (g). Finally, the chain of relations is used to infer the relationship of supplementary angles between A1 and A12 as shown in panel (h). This allows calculation of the measure of A12, the solution to the problem.

The sequence of steps in panels (d) through (g) illustrates a common problem-solving strategy, the use of subgoals. If a problem cannot be solved directly, the search through a person's knowledge store is aided by formulating subgoals that *can* be solved and that, once solved, contribute to solution of the original goal. In the present case, the general subgoal is finding a chain of quantitatively related angles. Specific subgoals are successively established as the distance between the goal angle and the currently known angle is reduced. Generating a subgoal can be thought of as establishing a new, intermediate representation of the problem. This serves to restructure or reformulate the problem so as to bring to light portions of it that *are* solvable, in the hope of working toward a full solution. In our earlier discussion of the processing requirements of word problems, we gave examples of different ways such problems could be reformulated to make their structure and limitations evident.

Subgoals can be viewed as a way of building a problem representation out of the givens of the situation, a way of encoding external task information for internal processing. This function of subgoals is most clearly seen in problems where different subgoals or representations produce very different search efforts. In the Magone (1977) two-string study, for example, subjects looked for anchor solutions, and this led them to evaluate the usefulness of objects quite differently. Larkin's (1977) studies comparing performances of novices and experts in the domain of physics found differences in the ways these groups routinely approached physics problems: Novices tended to attack a problem as given, that is, by immediately beginning to apply known principles and set up equations. Experts, in contrast, regularly began with a qualitative analysis of the problem, reformulating diagrammatic information in terms of a verbal interpretation before choosing solution strategies. One interpretation of such a finding is that experts are more skilled at setting soluble subgoals than are novices.

Greeno's (1978a) geometry investigations suggest that in actual problem-solving situations individuals typically generate the goals as they go along. Just as PERDIX does, they work on one goal at a time, starting with the goal closest to the solution state. If that goal cannot be met, they generate and try to solve a subgoal, and so on until they find a subproblem that is solvable [in the PERDIX example this happens in panel (e)]. Conceived in this way, it is apparent that goals and subgoals provide the comparisons between beginning, intermediate,

Analyzing Problems and Searching Knowledge Structures   225

and end states that motivate and direct the solution process. Thinking back to Polya's problem-solving hints in Chapter 6, it is easy to see how his strategies for promoting insight could be translated into strategies for generating subgoals in the context of a specific problem-solving task.

Rational and Empirical Analysis of a Problem-Solving Task

Having briefly outlined and illustrated a number of concepts that information-processing psychologists use to describe problem-solving strategies, we find it useful to point out the operation of these concepts in an actual problem-solving situation. Our choice for this explication is a series of experiments based on Wertheimer's parallelogram problem (see Fig. 6.3) (Resnick, Pellegrino, Morris, Schadler, Mulholland, Glaser, & Blumberg, 1980). We describe the basic experiment and several of its variations in detail because, taken together, they distinguish clearly between the role of specific conceptual knowledge and the role of more general problem solving strategies in enabling problem solution.

In these experiments, 10- and 11-year-old children were taught two separate procedures: They were taught how to compare the areas of two rectangles by exactly covering the figures with 1-inch square blocks and then counting the total number of blocks on each of the rectangles. They were also taught how to transform *non*rectangular paper figures into rectangular ones by cutting them with a scissors and replacing the pieces in a new way (and using all the pieces) to form a rectangle. These two processes are illustrated in Fig. 8.8. Following the training, children were presented with the problem of figuring out how to find the area of a parallelogram, a nonrectangular shape they had not seen in training.

Figure 8.9 suggests the kind of knowledge structures these children had probably built up by the end of their training on the measurement and transformation procedures. In the lower right segment of the diagram we see that they know a procedure for comparing the size of figures. They also know there is a limit on when to apply the procedure. They can only apply it when the figure to be measured is a rectangle (see arrows marked "allowable" object and "nonallowable" object). Connected to this *procedural* knowledge is *definitional* knowledge about the nature of rectangles and nonrectangles. Children possessing this segment of the knowledge structure, when presented with rectangles, should be able to measure them and to compare them for area. When confronted with a nonrectangle, these children would perhaps state that they could not apply the measurement procedure. Applying the criteria for well-structured knowledge, we might say the knowledge segment is integrated. It is also in good correspondence with the mathematical meaning of area (the block-placing and the counting procedure is an exact analog of the formula, area = height × width). But for solving the parallelogram area problem it is not adequately connected with other knowledge.

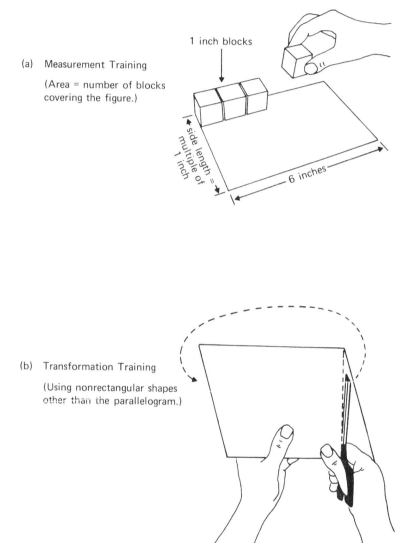

FIG. 8.8 Experimental version of the area-of-a-parallelogram problem. Solution requires transforming the parallelogram into a rectangle and then measuring by using 1-inch blocks.

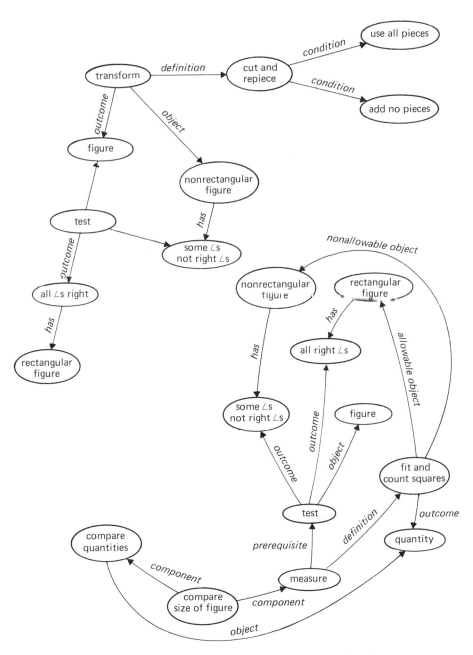

FIG. 8.9 Separate knowledge structures for comparing area and transforming figures.

228   8. INFORMATION PROCESSING ANALYSES OF UNDERSTANDING

The additional piece of knowledge needed is shown in the top left portion of the diagram. This is the knowledge structure that derives from having learned to transform nonrectangles into rectangles. As the figure shows, transformation training should result in a knowledge structure that has two important capabilities: It recognized when a figure is or is not a rectangle; and it knows that any plane figure can be cut up and reassembled as long as all pieces are used and none added (this is what preserves the equivalence of area under transformation).

The children in the experiments were taught these two separate procedures for measurement and transformation and thus presumably had in their minds something like the knowledge structures described in Fig. 8.9. But in the experimental task, after the two separate procedures had been taught, children were required to try to find for themselves the solution to the area-of-a-parallelogram problem, a solution that required them to build or "invent" the connection between the measurement and transformation knowledge structures. They were presented with two figures, at least one a nonrectangle, and were asked to find which was bigger.

Figure 8.10 shows the structure that we assume is needed in order to solve the parallelogram problem. It is a new structure formed when the knowledge concerned with area measurement and the knowledge concerned with transforming nonrectangles become connected. Once this connection is made, an individual faced with the problem of comparing areas of nonrectangles will have two knowledge networks to probe upon recognizing a nonrectangle. One network expresses the fact that the *measurement* procedure *cannot* be applied; the other expresses the fact that the *transformation* procedure *can* be applied. In this new knowledge structure, the outcome of the transformation procedure is represented as the production of a figure to which measurement can be applied. The person with this connected knowledge structure should be able to solve any area problem. The knowledge not only should apply to the parallelogram but should generalize to other nonrectangular figures as well.

How well did the 10-year-olds do in forming this connection between knowledge structures? Across the various experiments about 50% of the children succeeded in inventing the solution under the described conditions. A few immediately saw the connection between measurement and transformation. More, however, needed to do some active searching. Eventually, sometimes after several attempts to fit blocks on the parallelogram, they too realized that transformation (cutting) needed to come first. They did this without any explicit suggestion from the experimenter; we can thus say they invented the solution (i.e., built the connection needed in their knowledge structure) for themselves.

About 50% of the children, however, never did invent the connection by themselves. According to plan, the experimenters offered more and more explicit prompts if no solution appeared to be in sight after a child had been working for a while. When the experimenter eventually suggested it might be helpful to transform the parallelogram (i.e., use the scissors), every one of these noninventing

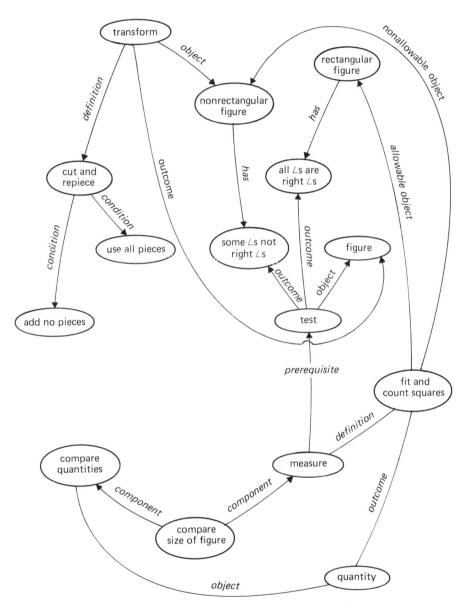

FIG. 8.10 Linked knowledge structures for comparing area and transforming figures.

230   8. INFORMATION PROCESSING ANALYSES OF UNDERSTANDING

children immediately cut the figures, transformed them into rectangles, and then went ahead with using the blocks for measurement. This shows unequivocally that they understood the two procedures separately and that all they needed was a "connection" to solve the problem. They were not able to produce the connection for themselves, but they were well able to use the hint supplied by the experimenter.

With this experiment in mind we can now review the effects of prior knowledge, the task environment, and strategy upon problem solving. With respect to prior knowledge, we see that having the requisite knowledge structures—established by training in transformation and blocks measurement—did not ensure that these structures would or could be used by all children in a problem-solving context. Indeed it was only by finding the connection between the two knowledge structures that children were able to find the area of the parallelogram. This "new" knowledge, the connecting link, had to be invented. On the other hand, for the invention to have taken place at all, the necessary knowledge segments had to be present. Invention and discovery did not happen in a vacuum.

But what exactly brought about invention in some children and not in others? To find an answer we must examine the details of the problem-solving situation and the protocols of experimental subjects under those conditions. Let us begin with a hypothesis about the goals children formulate when presented with the area-of-a-parallelogram task. Morris (1975) developed such a hypothesis in the form of a computer simulation. This minimodel of performance was used to analyze the task and develop the training routines that were supposed to build in the knowledge needed for invention. Figure 8.11 shows the task analyzed as a set of rules or condition–action statements, each of which has associated with it a routine for achieving the goal state. The rules can be interpreted as follows: To the left of each arrow is an activated goal and some reference to the stimulus condition (e.g., the presence of a certain kind of figure); to the right of each arrow is an action to be taken under the specified conditions.[2] The action can be either a physical one, such as transforming the figure, or it can be the setting of a subgoal, such as testing the applicability of the blocks routine. To aid the reader, the figure includes a verbal interpretation of each rule. Notice that rules FA (Find Area) 1 through FA 3 were explicitly taught during training, and the routines they call upon—Test Applicability, Use Blocks, and Transform—were also taught. In order to solve the problem successfully the students had to construct rule FA 4 for

---

[2]This process model is expressed as a "production system." As defined by Newell (1973), a production system is a scheme for specifying the hypothesized sequence of processing during task performance in the form of a series of action rules. When there is a match between objects or events in the task environment and the conditions (including an activated goal) specified in any condition–action rule, then it is assumed the appropriate action is executed. Each action results in a change in the stimulus or goal conditions, thus causing another action to be executed that is activated by the modified condition. In this way, a cycle of processing is initiated and continues until either there are no conditions corresponding to the available rules or all activated goals are met.

```
FA 1: ((GOAL: FIND. AREA) ──► LOOK AT. FIG)

        If you want to find how big a figure is, look at the figure.

FA 2: ((GOAL: FIND.AREA) AND (FIG) ──►(GOAL:
      TEST.APPLICABILITY.OF.BLOCKS))

        If you want to find how big a figure is, and you have a
        figure, then test to see if the blocks routine is applicable.

FA 3: ((GOAL: FIND.AREA) AND (YES.FIG) ──► USE.BLOCKS
      (SATISFY GOAL))

        If you want to find how big a figure is and it is a figure
        to which the blocks routine is applicable (i.e., a 'yes.fig')
        then use the blocks routine and the goal will be satisfied.

FA 4: ((GOAL: FIND.AREA) AND (NO.FIG) ──► (GOAL:
      TRANSFORM))

        If you want to find how big a figure is and it is a figure
        to which the blocks routine is not applicable (i.e., a 'no.fig')
        then try to transform the figure.
```

FIG. 8.11 A production system for finding the area of a figure. (From Resnick & Glaser, 1976. Copyright 1976 by Lawrence Erlbaum Associates, Inc. Reprinted by permission.)

themselves, and this constituted the invention. With this analysis of the task in mind, let us see if students' current goals can be inferred from their moment-to-moment behavior.

As mentioned earlier, students were allowed to pursue their chosen means of solution until they either solved the problem or reached a dead end. In the case of dead ends, the experimenter interjected a series of increasingly explicit prompts that eventually cued the proper solution strategy. A look at the behavior of eventual inventors (those who did not solve immediately) in response to the initial prompt reveals an interesting difference in their apparent goals. For the slower inventors—those who needed to do some active searching before seeing the connection between measurement and transformation—the problem-solving session typically proceeded as follows: They would begin by filling the rectangle with blocks, this being a "legal" move within the measurement knowledge structure. Then they would try putting blocks on the parallelogram, often expressing considerable uneasiness as they did so. After a while the experimenter would say "That's wrong." She never said *what* was wrong, but these inventors-to-be apparently had a ready explanation: It was wrong to be putting the blocks on a figure of that particular shape. How do we know this? Because these children would immediately clear off all the blocks and go back to pondering the parallelogram.

The noninventors began much like the others, first filling the rectangle with blocks and then going on to the nonrectangle. Like a number of the inventors, they worked at filling up the nonrectangle, sometimes even expressing concern about the blocks not fitting. Then came the experimenter's admonition, "That's wrong." These children, too, had an interpretation of what was wrong, but their interpretation seemed to be very different from the inventors'. Unlike the eventual inventors, they did not wipe the blocks off the figure entirely but tried instead to rearrange them, perhaps lining them up along the diagonal instead of the base or attempting to "squeeze" them into corners. The noninventors persisted in this rearranging; not one of them spontaneously removed the blocks and reconsidered the problem with a clean slate.

Based on these observations of behavior, we can infer that inventors and noninventors were working on different goals when interrupted by the experimenter's prompt. They each seemed to interpret the prompt in terms of their own current aims. One group behaved as if they were testing to see if the blocks would fit on the parallelogram, and they treated the experimenter's warning as information that the blocks would not fit. The other group behaved as if they were trying to measure the parallelogram and treated the prompt as information that they had fit the blocks on the wrong way. The group concerned with the applicability of blocks was able to rethink the problem by clearing away the blocks and looking again at the blank figure, thus allowing the nonrectangular shape to cue the transformation routine (FA 4). Another way of saying this is that each child's *representation* of the immediate subproblem—as either testing for applicability of the blocks routine or fitting the blocks on the figure—determined how new information would be interpreted. Each problem solver's representation, then, was apparently crucial in determining whether or not he or she solved the problem without explicit hints.

In subsequent experiments the task-environment and the goal-setting process were deliberately manipulated in the hope of determining which aspects of the test situation were facilitating invention. The same general training scheme was used, but the task conditions were systematically varied. Across the experiments, there was some variation in the figures presented during the problem-solving session. Figure 8.12 shows the possibilities. There might be one or two figures, a rectangle might or might not be included, and the nonrectangle might be either a parallelogram or an analogous figure in which the "gap–extra" relationship was especially evident. In general, the absence of a rectangle or the presence of a gap–extra-type figure produced more solutions. The gap-extra figure probably cued the transformation procedure directly, an interpretation supported by the frequency and primacy with which transformation was mentioned during overt goal analysis in later experiments. Analysis of the sequence of behaviors during problem solving suggests that the presence of the rectangle led most subjects to immediately employ the blocks-measurement routine. For many children, once they had measured one of the figures, it was difficult to switch goals, and they

Analyzing Problems and Searching Knowledge Structures   233

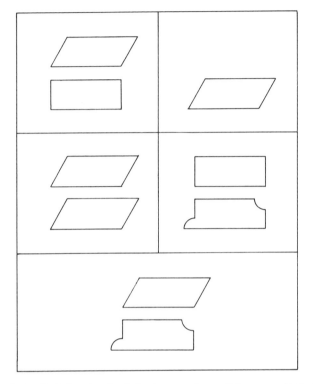

FIG. 8.12 Variations in stimuli for the area-of-a-parallelogram problem. Different sets of figures cued different learned routines, leading to different problem-solving strategies.

persisted—as we have already seen—in trying to place blocks on the parallelogram. In these cases, characteristics of the task environment prompted specific solution strategies, apparently because they led to different representations of the problem.

Another variation on the parallelogram experiment showed the power of explicitly stated goals in building problem representations and determining solution strategies. Again, children were given the transformation and measurement training and then were presented with two figures, at least one a nonrectangle. This time, however, instead of asking the children to find which figure was bigger, the experimenter said nothing except, "What do you think I want you to do?" Each answer was recorded and the question was repeated until no new ideas were forthcoming. All the children thought of both measuring *and* transforming, and many thought of testing explicitly to see whether the figure was a rectangle. In other words, they formulated alternative goals in the course of "guessing" what the experimenter wanted. One might expect that this formulation in advance would lead them to recognize the connections between the measurement and transformation knowledge structures. And indeed, that is what happened, for when finally asked to find which figure was bigger, virtually every child went on to invent a solution. This suggests that overt attention to the possible goals in a problem-solving situation can dramatically improve the likelihood of discovering the solution. It is only a hint at the power of conscious analysis of goals, but the empirical data fit well with the speculative accounts of both mathematicians and psychologists.

Facilitating Problem Solving Through Instruction

What are the possibilities for improving the likelihood of problem solving and invention? Using the three basic ingredients of problem solving that we have discussed, namely, prior knowledge, the task environment, and strategy, we can suggest instructional interventions that might improve the functioning of each component in the problem-solving situation. First, instruction can try to ensure the presence of well-structured knowledge and to maximize the links to related concepts and procedures. This means making sure mathematics instruction is mathematically honest, reveals the subject-matter structure, and provides the opportunity to practice new procedures and concepts in a wide variety of contexts. This also means teaching as much mathematical knowledge as possible, given the age and ability of the student. From a purely intuitive standpoint, the more facts, procedures, and relations that characterize a person's knowledge structure, the more likely that person is to invent or discover needed connections. If prerequisite knowledge is lacking in a student, we cannot expect skilled problem solving.

Instruction can also try to take into account the role of the task environment as the primary stimulus for problem solving. To give practice in problem-solving strategy, for example, one can state the problem clearly, with a minimum of extraneous information, or one can accompany the problem statement with pictures and diagrams. Children might be taught to scan features of the environment, allowing their mind free play to consider the connections each feature calls up. Or they might learn to write down their thoughts or draw pictures so as to create for themselves a richer task environment out of which to build a problem representation.

Finally, it may be possible to design instruction in specific problem-solving strategies. The parallelogram experiments, for example, give strong support to the notion of teaching explicit goal analysis, that is, teaching children to think ahead and visualize many courses of action, and their consequences, before launching into solution procedures. The idea of teaching generalized problem-solving skills has been given extensive consideration by Polya (1945/1957) and Wickelgren (1974). Recall Polya's questioning strategy: "What is the unknown? What are the data? What is the condition?" Such questioning may have the effect of highlighting both the features of the task environment and the goals of the problem, thereby contributing to the call-up of relevant knowledge and selection of appropriate solution strategies. Wickelgren (1974) has developed a problem-solving "bag of tricks" such as working backward from the solution, setting subgoals, and evaluating progress at various points during a heuristic search. The possibilities for this type of instruction have only begun to be thought of, much less tested and perfected. But empirical work on mathematical problem solving such as we have described holds promise for the kind of understanding we need in order to create better problem solvers.

SUMMARY

In this chapter we have considered the various ways in which modern information-processing psychology is attempting to deal with questions of understanding and structural knowledge in mathematical learning and thinking. Our concern is to find systematic ways of representing knowledge that also account for the human capacity to reason beyond the literally given information, to find connections and relationships among different areas of knowledge, and to apply knowledge in problem-solving situations.

A first task in developing a systematic psychology of understanding is to represent people's flexibility in thinking and reasoning. Current theories of semantic (i.e., long-term) memory, particularly network models of knowledge, represent the interrelationships—and thus the structure—in what people know. In these models, ideas and concepts are viewed as standing in specific relationship to one another, and it is explicitly understood that learning may consist of constructing new interconnections and relations as well as receiving new items of information. A knowledge structure for arithmetic operations was used to illustrate and elaborate these points. We proposed that well-structured knowledge about mathematics could be viewed as a primary goal of mathematics instruction. Three criteria for well-structuredness were outlined: correspondence (the match of one's mental picture with correct mathematical concepts), integration (the degree of interrelatedness of concepts within a particular domain of mathematics), and connectedness (the degree to which knowledge in one domain of mathematics is related to knowledge of other domains). Examples of research in physics demonstrated changes in correspondence and integration as people progressed in their learning of the subject matter. The behavior of a young girl on subtraction problems revealed changes in connectedness and integration that came about through instruction.

Next we considered the question of how knowledge is used in solving problems. We introduced the notion of problem-solving strategies that call upon relevant knowledge structures and help direct the search for needed information. Several interrelated aspects of problem-solving strategy merit attention: problem representation, the role of the task environment, and the impact of different strategies. The mental representation of a problem plays an important role in directing the likely course of solution attempts, and both the task environment and individual differences in knowledge structure influence the particular problem representation that is achieved. Experiments on discovery learning and functional fixedness show that verbal instructions and physical aspects of the task environment are especially influential in cuing specific initial responses. Once an initial problem representation is established, a number of strategies can come into play. Among these are generate-and-test strategies, heuristic search, inference, and the setting of subgoals. Examples of problem-solving strategy were seen in the context of a computer program that solved geometry problems.

A series of experiments, all based on a variant of Wertheimer's parallelogram problem, illustrated the use of knowledge and strategy in problem solving. In these experiments, children were taught a procedure for finding the area of rectangles and a procedure for transforming nonrectangles into rectangles. They then had to invent a procedure for finding the area of a parallelogram. The experiments demonstrate that even secure knowledge of the component skills does not ensure successful problem solution. The child's particular representation of the problem (the goal) at the time the experimenter provided new information affected the way in which the information was interpreted, and this in turn affected the likelihood of problem solution. In addition, the shape of the figures presented seemed to have an impact on the rate of solution, in that some figures cued a gap-extra interpretation whereas others invited attempts to apply the area procedure directly. A demand that goals be analyzed in advance was particularly effective in prompting successful solution strategies. Based on these findings, we suggest that specific problem-solving skills such as explicit goal analysis might be taught and that instruction should strive both to ensure well-structured knowledge and to take into account the role of the task environment in learning and problem solving.

REFERENCES

Anderson, J. R. *Language, memory, and thought.* Hillsdale, N.J.: Lawrence Erlbaum Associates, 1976.

Anderson, J. R., & Bower, G. H. *Human associative memory.* New York: Halsted Press, 1973.

Brown, J. S., & Burton, R. R. Diagnostic models for procedural bugs in basic mathematical skills. *Cognitive Science,* 1978, *2,* 155-192.

Case, R. Piaget and beyond: Toward a developmentally based theory and technology of instruction. In R. Glaser (Ed.), *Advances in instructional psychology* (Vol. 1). Hillsdale, NJ: Lawrence Erlbaum Associates, 1978.

Chase, W. G., & Simon, H. A. The mind's eye in chess. In W. G. Chase (Ed.), *Visual information processing.* New York: Academic Press, 1973.

Collins, A. M., & Quillian, M. R. Retrieval time from semantic memory. *Journal of Verbal Learning and Verbal Behavior,* 1969, *8,* 240-247.

Collins, A. M., & Quillian, M. R. How to make a language user. In E. Tulving & W. Donaldson (Eds.), *Organization of memory.* New York: Academic Press, 1972.

Gagné, R. M., & Brown, L. T. Some factors in the programing of conceptual learning. *Journal of Experimental Psychology,* 1961, *62*(4), 313-321.

Geeslin, W. E. *An exploratory analysis of content structure and cognitive structure in the context of a mathematics instructional unit.* Unpublished doctoral dissertation, Stanford University, 1973.

Geeslin, W. E. *Comparison of content structure and cognitive structure in the learning of probability.* Paper presented at the meeting of the American Educational Research Association, Chicago, April 1974.

Greeno, J. G. Cognitive objectives of instruction: Theory of knowledge for solving problems and answering questions. In D. Klahr (Ed.), *Cognition and instruction.* Hillsdale, N.J.: Lawrence Erlbaum Associates, 1976.

Greeno, J. G. A study of problem solving. In R. Glaser (Ed.), *Advances in instructional psychology* (Vol. 1). Hillsdale, N.J.: Lawrence Erlbaum Associates, 1978. (a)

Greeno, J. G. Understanding and procedural knowledge in mathematics education. *Educational Psychologist,* 1978, *12*(3), 262-283. (b)

Katona, G. *Organizing and memorizing.* New York: Columbia University Press, 1940.

Larkin, J. H. *Problem solving in physics.* Working paper, University of California, Berkeley, Group in Science and Mathematics Education and Department of Physics, 1977.

Magone, M. E. *Goal analysis and feature detection as processes in the solution of an insight problem.* Unpublished masters thesis, University of Pittsburgh, 1977.

Maier, N. R. F. Reasoning in humans I: On direction. *Journal of Comparative Psychology,* 1930, *10,* 115-143.

Morris, L. L. *An information processing examination of skill assembly in problem solving and development using Wertheimer's area of a parallelogram problem.* Unpublished doctoral dissertation, University of Pittsburgh, 1975.

Newell, A. Production system: Models of control structures. In W. G. Chase (Ed.), *Visual information processing.* New York: Academic Press, 1973.

Newell, A., & Simon, H. A. *Human problem solving.* Englewood Cliffs, N.J.: Prentice-Hall, 1972.

Norman, D. A., & Rumelhart, D. E. *Explorations in cognition.* San Francisco: W. H. Freeman, 1975.

Polya, G. *How to solve it* (2nd ed.). New York: Doubleday Anchor Books, 1957. (Originally published 1945.)

Reitman, J. S. Skilled perception in Go: Deducing memory structures from inter-response times. *Cognitive Psychology,* 1976, *8*(3), 336-356.

Resnick, L. B., & Glaser, R. Problem solving and intelligence. In L. B. Resnick (Ed.), *The nature of intelligence.* Hillsdale, N.J.: Lawrence Erlbaum Associates, 1976.

Resnick, L. B., Pellegrino, J. W., Morris, L. L., Schadler, M., Mulholland, T., Glaser, R., & Blumberg, P. *Wertheimer revisited: Information-processing analyses of invention in the area-of-a-parallelogram task.* University of Pittsburgh, Learning Research and Development Center, 1980.

Rothkopf, E. A., & Thurner, R. D. Effects of written instructional material on the statistical structure of test essays. *Journal of Educational Psychology,* 1970, *61,* 83-89.

Shavelson, R. J. Some aspects of the correspondence between content structure and cognitive structure in physics instruction. *Journal of Educational Psychology,* 1972, *63*(3), 225-234.

Shavelson, R. J. Methods for examining representations of a subject-matter structure in a student's memory. *Journal of Research in Science Teaching,* 1974, *11*(3), 231-249.

Shavelson, R. J., & Stanton, G. C. Construct validation: Methodology and application to three measures of cognitive structure. *Journal of Educational Measurement,* 1975, *12*(2), 67-85.

Thro, M. P. Relationship between associative and content structure of physics concepts. *Journal of Educational Psychology,* 1978, *70*(6), 971-978.

Wertheimer, M. *Productive thinking* (Enlarged ed.). New York: Harper & Row, 1959. (Original edition published 1945.)

Wickelgren, W. A. *How to solve problems: Elements of a theory of problems and problem solving.* San Francisco: W. H. Freeman & Co., 1974.

CONCLUSION

9 Looking Ahead

The psychology of mathematics is changing rapidly. It is abundantly clear that by the time this book is read there will be important new developments that we cannot, of course, have reported or commented on. Despite this sense of having written an incomplete story, there are important reasons to highlight the current state of this particular topic in psychology. What we have done in the course of this book is to review and interpret an important segment of an emergent *psychology of instruction*. Instructional psychology is growing in strength and visibility as psychologists increasingly turn their attention to how people acquire and use information in complex domains of knowledge.

For a number of reasons, the domain of mathematics is probably an "advance guard" of the instructional psychology of the future. As a school subject, and even as an area for scholarly investigation, mathematics is more clearly bounded and well defined than almost any other instructional domain. Mathematicians provide the psychologist with formalized statements of the subject matter, thus simplifying the process of rational task analysis. Further, the subject matter of mathematics is largely self-contained; compared even with physics there is less dependence on information external to the formal domain. We thus need to search less widely for the knowledge structures likely to be relevant to mathematics learning. Finally, of all school subject matters, mathematics is the least dependent on the processing of natural language. This means that the psychology of mathematics can advance substantially with only limited attention to the complexities and difficulties of language processing. It seems reasonable to suppose, therefore, that as more psychologists turn their attention to this area of study, the psychology of mathematics can relatively quickly provide the kinds of accounts of subject-matter learning that will eventually characterize much of instructional psychology. In this concluding chapter, we outline what we see as some of the important questions the psychology of mathematics instruction is likely to address and how these questions are central to a more general instructional psychology. We begin by considering briefly the changing nature of psychologists' concern with questions of instruction.

THE CHANGING NATURE OF INSTRUCTIONAL PSYCHOLOGY

The chapters of this book trace a historical progression both in the involvement of psychologists with instructional questions and in how psychologists have attempted to understand and study these questions. For Thorndike and his contemporaries, instruction was a proper domain of study for the scientific psychologist. Psychologists did more than offer general principles of learning and instruction to educators; they themselves became involved in the process of interpretation and development that made direct applications to instruction possible. Thorndike moved easily between theoretical and practical work. As we have shown in Chapter 2, he did more than suggest that school subjects could be analyzed in terms of bonds so that general principles of association could be applied to strengthening arithmetic knowledge. He actually performed the analyses himself and even participated in developing a set of arithmetic textbooks (Thorndike, 1917–1924) that embodied his principles of instruction. Thorndike's psychology of instruction, then, was subject-matter specific and yet clearly rooted in a general theory of learning. Thorndike recognized that only by carefully analyzing the subject matter to be taught could the theory of learning be applied. His analyses of arithmetic in terms of bonds were, in a sense, instructional psychology's first task analyses.

Although Thorndike's influence never entirely disappeared, research on arithmetic learning, as on other parts of the school curriculum, became for several decades in the middle of this century the almost exclusive province of educational researchers who were neither trained in nor influential in the psychological research community. With a few exceptions, such as Brownell, the language of psychology began to drop out of these people's work, and questions of immediate application (as in organizing drill programs) dominated to such a degree that general questions of how learning took place or why certain problems should be more difficult than others were barely raised. Work of this kind neither effectively drew from nor contributed to the mainstream of psychological research.

Meanwhile, experimental psychologists were attempting to create a "pure" science, uncontaminated by concerns with educational applications. Both the subject-matter orientation and the easy flow between practical and theoretical concerns that had characterized Thorndike's work disappeared from psychology for some time. It is not accidental that none of the major theorists of learning after Thorndike have been mentioned in this book or that we have made no effort to explore the instructional implications of Hullian learning theory, for example, or of Tolman's more "cognitive" theory. The absence of these psychologists from these pages reflects the fact that neither the major learning theorists themselves nor their students seriously attempted to explore such connections. The separation of scientific psychology from research on instruction was nearly complete.

After World War II, spurred perhaps by their involvement in many practical efforts during the war, psychologists began to turn once again to questions of instruction. Two lines of investigation could be delineated, one based on principles of behavioral learning psychology and one on the emergent American cognitive psychology. Among behavioral psychologists, B. F. Skinner was the leader in this renewed focus on instruction. Yet Skinner's work, important as it was in drawing the attention of many psychologists to instructional questions, has not been discussed in this book. That is because his prescriptions for instructional design were general rather than specific; applications to any specific domain of learning were not obvious. Out of Skinner's principles—positive reinforcement, errorless learning, and successive approximations—grew the programmed instruction movement. But beyond the prescription for stating instructional objectives in terms of overt performance, there was no theory of subject matter embedded in Skinner's work and no clear suggestion for how to analyze a domain of knowledge to make it more tractable for systematic instruction.

It was really Gagné, with his theory of cumulative learning and his procedures for hierarchical analysis of subject-matter tasks, who returned a systematic subject-matter concern to behavioral learning psychology. Although Gagné's own analysis of mathematics was limited to a few tasks chosen as examples for demonstrations and experiments, others used the general theory of hierarchy analysis to analyze extensive segments of the mathematics curriculum and to build instructional programs based on their analyses. Gagné's initial work was entirely behavioral in orientation and sought to show only that knowledge acquisition was organized hierarchically. There was no effort to explain *why* tasks ordered themselves into particular hierarchies of performance or learning. But, as we showed in Chapter 3, others building on Gagné's work began to introduce a form of cognitive process analysis into the building of hierarchies. These process analysts attended to the flow of behavior during actual performance of mathematical tasks and offered explanations of task difficulty in terms of memory and perceptual processing demands. Although the resulting hierarchy/process analyses were rational analyses, they were able to make contact with the emerging research in empirical task analysis rooted in cognitive information-processing theory, as described in Chapter 4.

This represents one point of jointure between learning and cognitive psychology. For the most part, however, cognitive and learning psychologists proceeded along quite separate paths in pursuing instructional questions. Bruner's involvement with mathematicians in redesigning the elementary school mathematics curriculum, described in Chapter 5, was typical of cognitive psychology at the time. The postwar American cognitive psychologists were concerned with *what* people knew and understood; it was explicitly the *content* of a subject matter that interested them—how it was structured, and how instruction might be organized to help children learn the subject-matter structures that would let them think mathematically. Renewed attention to European traditions of research on human thinking, including the work of both Piaget and the gestalt psychologists, confirmed and sharpened a focus among many American cognitive psychologists on how the subject matter of mathematics was understood and on how bodies of information were mentally ordered so as to permit reasoning and problem solving in mathematical tasks. On the practical side, these concerns suggested redesigning the content of the mathematics curriculum so as to stress structural understanding, and developing ways to represent fundamental mathematical concepts so as to make them accessible to children. More recently, information-processing psychologists have turned their attention to the issue of understanding in mathematics, and they are now studying it in a form that links performance to understanding rather than pitting them against one another as competing goals for instruction.

NEEDED: A COGNITIVE LEARNING PSYCHOLOGY

This brief history of psychologists' changing interest in subject-matter instruction highlights a curious fact. The old learning psychology had much to say about how to arrange conditions for learning; but it was weak in its ability to describe the content of learning. Cognitive psychology, by contrast, offers rich descriptions of the content and processes of performance in a subject matter. But up to now, it has said almost nothing about how competence is acquired.

Cognitive psychology has been largely concerned with describing "moments" of performance or "states" of understanding. It can be characterized as a "transparent snapshot" psychology, in which mental processes are depicted at a given point in time. An underlying assumption is that no important changes in processes take place during the period being studied and modeled. This general characterization is as true of developmental studies of cognition as it is of studies confined to a single age group. Developmental studies typically compare the performances and underlying knowledge of people of different ages. But the descriptions for any one age are snapshot descriptions. In short, cognitive psychology has shown no serious interest, until very recently, in trying to account for *changes* in capability. There has been no strong theory of how cognitive knowledge or skill is *acquired* and no theory of *transitions* in competence.

Traditional learning theories, on the other hand, contrast with cognitive theories in two respects. They are "opaque," seeking to describe only overt behavior and not its mental basis. But they record "movies" rather than snapshots of human behavior. Methods for studying changes in performance over time are highly developed and transitions in competence are a central concern. The assumed opacity of mental processes, however, makes it hard for learning psychology to deal with intellectual performances such as are involved in mathematics. Thorndike attempted a form of task analysis, but mathematics instruction based on his theory had to depend in the end on purely empirical orderings of tasks. It could not effectively define the kinds of knowledge and understanding that might underlie computational competence. Skinnerian theory, explicit in some of its prescriptions for how to organize teaching, rejected the very notion of mental processes, such as knowing or understanding, and thus treated the problem of defining subject matter as if it were a minor preliminary to the real task of instruction.

For those interested in building a psychology of instruction, this situation has been discouraging. The richness of psychological description that characterizes cognitive psychology has made it abundantly clear that we need not continue to follow a content-empty behavioral approach to instructional analysis. On the other hand, cognitive psychology's focus on essentially static descriptions of performance and competence has meant that the implications of cognitive task analyses for instruction are often difficult to detect. The sustaining hope has been that a more thorough knowledge of how people understand and perform tasks in a subject-matter domain will provide the "targets" toward which instruction should be directed—the "cognitive objectives" for instruction. But it has not been clear exactly how to make the connection between complex descriptions of "moments" of performance and detailed specifications for organizing instruction.

If practice-oriented readers of this book—mathematics educators or curriculum designers perhaps—have found themselves impatient at times with the apparent distance between present psychological concerns in the psychology of mathematics and their own concerns about the nature and course of instruction, this is probably why. Reflecting the state of the field, we have been able to report largely static descriptions of performance and understanding in mathematics. These can be viewed as a contribution to the psychological definition of the mathematics curriculum. But we could neither raise nor begin to answer, in a form well grounded in the psychology available to us, questions about how to foster transitions in competence. What instructional psychology needs is "transparent movies." Instructional processes cannot be greatly clarified and improved unless we can *both* look into the mind *and* chart changes in capability. Only then can we develop theories about how to present information and organize exercises in ways that will help students to acquire and strengthen mathematical skill and knowledge.

Yet cognitive psychology appears to be changing in ways that will make it more useful for instructional theory and practice. A number of cognitive psychologists have expressly recognized the need to reintroduce questions of transition and acquisition into their work. A few have begun actually to study the cognition of learning (see, for example, Anderson, in press), and a new cognitive learning psychology seems to be on the horizon. The future of the instructional psychology of mathematics lies squarely within this emergent cognitive learning psychology. In fact, it seems likely that the new cognitive learning psychology can only develop strongly to the extent that detailed studies are undertaken on the learning of particular subject matters. The questions addressed and the problems reformulated as the psychology of mathematics proceeds can thus be expected to help shape the new cognitive psychology of learning.

QUESTIONS FOR THE PSYCHOLOGY OF MATHEMATICS INSTRUCTION

Trying to predict the future of scientific developments is risky. Nevertheless, there are some emerging questions and themes in the psychology of mathematics instruction that can be identified now and that are likely to become increasingly important over the next decade. A consideration of these gives a sense of the possibilities and directions for a new, cognitively grounded instructional psychology.

Computational Skill and Mathematical Understanding

The relationship between computational skill and mathematical understanding is one of the oldest concerns in the psychology of mathematics. It is also one that has consistently eluded successful formulation as a research question. Over the years, the issue has been posed in a manner that made it unlikely that fruitful research could be carried out. Instead of focusing on the *interaction* between computation and understanding, between practice and insight, psychologists and mathematics educators have been busy trying to demonstrate the superiority of one over the other. Some psychologists, like Thorndike, believed that computational skill preceded understanding; others, such as Bruner and the gestalt psychologists, believed that understanding provided the basis for developing computational skill. On both sides, argument proceeded more by assertion and demonstration than by scientific investigation. Experiments could be designed to show the apparent benefit of either computational practice or "meaningful" learning. But as we have tried to show in analyzing some of these experiments, the different methods of teaching each seemed to produce whichever kind of knowledge the lesson itself stressed. The *relationships* between skill and understanding were never effectively elucidated.

What is needed, and what now seems a possible research agenda, is to focus on *how* understanding influences the acquisition of computational routines and, conversely, on how, with extended practice in a computational skill, an individual's mathematical understanding may be modified. In the course of this book we have described several studies that indicate changes over time in the performance strategies for arithmetic operations and other mathematical procedures. In these studies, the only known intervening "instructional" event was practice on the class of tasks under study. In the Groen and Resnick study (see Chapter 4), for example, there was no attempt to explain the logic of addition beyond the initial sessions nor especially to point out features of the task that might suggest new procedures or strategies. Practice alone produced a form of learning in which the ability to perform a class of calculations was strengthened, and at the same time a more efficient procedure was developed by the learners themselves.

This research offers a hint of what may be a very general phenomenon in mathematics learning: reorganization of procedures by the learner in the course of computational practice. Other studies have begun to appear that describe transformations in problem-solving strategies in the course of practice on classes of problems (e.g., Neches & Hayes, 1978), and we can expect to see more such studies, at least some of them focused on mathematical tasks. But the careful mapping of transitions, both in procedures and in the knowledge that underlies them, is still largely a task for the future. Once these mappings have been made for a number of important mathematics tasks, it will be possible to turn our attention to the processes involved in actually making the transitions. The psychology of mathematics can then deal with learning not simply as a function of practice but as a function of the kinds of mental processes that people engage in as they practice, processes that permit them to modify their understanding of mathematical concepts. At the same time, carefully charting the changes in knowledge structures as competence develops should allow us to discover the ways in which mathematical understanding may influence the development of efficient computational performance.

The Role of Practice in Learning

What do people do when they practice, other than going through the initially learned procedure over and over again? It has been suggested (Klahr & Wallace, 1976) that practice serves to "automatize" the components of a procedure so that there is more "space" in working memory for scanning the task environment to detect regularities that might be incorporated into shortcut procedures. Which regularities are the most likely to be noticed, and how does the form in which the initial procedure is taught affect what is noticed? Are there ways of managing practice to enhance the likelihood that invention of shortcut procedures will result?

Research that addresses practice in these terms will be returning to one of the oldest questions in the psychology of mathematics. But this research will ask not only how practice builds efficient and reliable performance in calculation but also how the building of such skill is related to the development of understanding. We will want to know, as a result of this research, how to organize practice to maximize learning. But we are not so likely to argue, as some successors to Thorndike did, for making decisions simply on the basis of which problems are the empirically more difficult. Instead, we want to know how particular sequences of practice problems may guide attention in ways that modify performance strategies and the conceptual knowledge on which they are based.

An important part of the methodology for studying the effects of practice is already in hand in the form of latency studies. With only a few exceptions, latency studies of arithmetic performance up to now have been "snapshot" studies, not attempting to plot changes in knowledge and strategy over time. But the same methodologies, used in longitudinal research designs in which individuals' performances are tracked over relatively extended periods of practice (perhaps several months or longer), ought to allow us to collect data on how strategies are modified by practice and by the acquisition of new information. Combined with interviews and other direct task analysis methods, these kinds of longitudinal studies promise to offer us a *cognitive* psychology of drill and practice. This psychology will gauge the effects of practice on knowledge structures and on the actual procedures used in performance rather than simply relating practice to the likelihood of achieving correct answers.

Mental Representations in Learning

Research on the relationship between skill and understanding will carry with it the question of how quantity, operations on quantity, and related mathematical concepts are represented mentally in the course of learning and performance. A special aspect of this question is the role of physical "metaphors" for mathematical concepts, that is, physical representations that allow overt manipulation of materials in ways that help link performance algorithms to their underlying mathematical principles. A concern with representations of mathematical concepts was raised both by gestalt psychologists, with their emphasis on "good" and "sensible" structuring of mathematical ideas as a basis for problem solving, and by American cognitive psychologists, such as Bruner, who stressed the importance of simple but mathematically correct representations as the basis for initial learning of mathematical concepts. As described in Chapter 5, the mathematics teaching reform movement of the 1960s centered in part on the problem of how to provide simplified, often concrete representations of complex mathematical concepts so as to make these concepts accessible to children. For example, a variety of structural mathematics materials such as Dienes blocks or the Montessori materials could be used to represent base structures in the number system. It was assumed that by working with the physical materials children would build up mental representations of the concepts, and these mental representations would allow them eventually to solve related problems without the physical aids.

Although plausible enough, this assumption has never really been tested. There exists no psychological theory of what constitutes a useful physical representation of a mathematical concept. It is widely assumed, for example, that performing addition and subtraction using Dienes blocks, with block exchanges representing processes of carrying and borrowing, will build mental representations of computational procedures that map directly to the procedures used when no concrete materials are available. But neither the process of mapping nor the nature of the mental representation that is formed through the use of such materials has been directly studied. No one has yet attempted to probe the content of mental representations induced by materials such as the Dienes blocks. Nor has anyone identified the processes by which these become connected with representations that underlie pencil-and-paper calculation algorithms. It is possible, for example, that the blocks are used by many children as if they are simply a calculating machine: The processes these children engage in are a way of arriving at the answer to a problem, but the manipulations may not be seen as "representing" the operations performed in pencil-and-paper addition and subtraction. The physical and the pencil-and-paper procedures, then, may exist side by side but unconnected in a child's mind. In such a case the concrete representation does little to inform initial learning or understanding of the more "abstract" procedure. Alternatively, the physical materials may function as a convenient way to explain or justify why the pencil-and-paper procedure works *after* a child already knows the pencil-and-paper procedure, but they may not be an effective means of learning the procedure in the first place.

The use of concrete instructional materials, then, poses several fundamental questions for the new cognitive learning psychology of mathematics: What is the nature of the knowledge structures induced by working with concrete structure-oriented materials? What is the nature of the knowledge structures developed in the course of practice with the more abstract (mental or pencil-and-paper) procedures? How do these knowledge structures map onto each other? How do they become connected and how can connections be induced through particular forms of instruction or practice? Finally, when connections are made, do they enhance performance of the abstract routines, and do they increase the likelihood of generalizing to related calculation routines or of inventing new routines?

Constructing Mathematical Knowledge Versus Being Told

One of the fundamental assumptions of cognitive learning psychology is that new knowledge is in large part "constructed" by the learner. Learners do not simply add new information to their store of knowledge. Instead, they must connect the new information to already established knowledge structures and construct new relationships among those structures. This process of building new relationships is essential to learning. It means that mathematical knowledge—both the procedural knowledge of how to carry out mathematical manipulations and the conceptual knowledge of mathematical concepts and relationships—is always at least partly "invented" by each individual learner. Although this conceptualization of mathematics learning seems to flow naturally from today's cognitive psychology, it is noteworthy that virtually no research has yet been done on the processes of invention in mathematics. Older work on discovery in mathematics learning (see Chapter 6) was largely an attempt to show that the opportunity to induce principles was more conducive to retention and transfer of information than was rote drill. As in the case of computation and understanding, practice and discovery were pitted against each other instead of being treated as complementary aspects of a constructive learning process.

A few studies on the role of inference in the development of knowledge structures now exist (see Chapter 8), but these represent only a small start at investigating the complex relationship between being told or shown mathematical information and constructing procedures that use this information and link it to existing knowledge structures. Research is required on the kinds of knowledge structures that are conducive to the inventions that characterize learning. What are the processes of invention themselves like? Is there a "moment of insight" in the sense that the gestalt psychologists appeared to suggest or is the course of learning more gradual? Is it perhaps a set of accretions in which very small changes in performance and knowledge structures cumulate into what is in effect a new cognitive structure, much as a new structure was built by PERDIX through step-by-step procedures of inference in the geometry example of Chapter 8?

A particularly important question is what happens when new information enters the knowledge system. What are the processes by which what we are told or shown becomes integrated into established knowledge structures? In the old associationism of Thorndike's time new information would simply be added to the list of bonds already in one's store of knowledge. The instructional question was how much practice was required under what conditions to have new knowledge enter one's store and to strengthen it. In the new cognitive learning psychology our questions will be posed differently, but there must continue to be a concern with how new information, presented by others, enters the individual's knowledge structure. How does it become integrated? What connections are made? Is it possible to tailor the form of information to an individual's existing knowledge so as to make the process of connection building easier? Finally, what kind of rehearsal or practice is needed for these constructions to take place?

Individual Differences

Research on practice, on understanding, indeed on all the questions outlined in this chapter will undoubtedly reveal important differences among individuals in the character of processes observed. Psychologists studying learning and performance in any subject matter always encounter different responses to the same task on the part of individuals. This is true even when they attempt to select a population for study that is homogeneous with respect to general learning abilities and prior exposure to the relevant curriculum. Unless the research has been designed specifically to study individual differences, however, there is usually no way to account for the differences when they appear in the data. Should the differences be attributed to stable individual characteristics, or *traits*, or are they a result of an individual's momentary state of knowledge and practice with respect to a given task?

Both trait and knowledge state explanations for individual differences can be found in instructional theory and research. Trait explanations are assumed when it is proposed that instruction be matched to individuals' aptitudes. Research on aptitude-treatment interactions (ATIs), for example, generally assumes that certain ways of approaching cognitive learning tasks are stable individual traits that will recur from task to task. ATI research attempts to determine whether there are different general approaches to instruction that will be optimal for individuals with different sets of characteristics. An extensive review of the ATI literature (Cronbach & Snow, 1977) has found few consistent patterns of interaction. Thus, although the idea of matching instruction to aptitudes continues to be appealing, there is little scientific basis for such matching at the present time.

Knowledge state explanations of individual differences are implicit in hierarchy approaches to instructional design and research. The hierarchy approaches assume that if we could completely identify individuals' knowledge and skills as they began to learn a new task, we would find that all individuals in a given knowledge state would respond to a particular task in the same way. When hierarchies of learning objectives are used as a way of individualizing progress through the mathematics curriculum (see Chapter 3), it is assumed that if the necessary prerequisite knowledge is taught before a new task is encountered, then individual differences in learning the new task should be sharply reduced. As a general observation, this reduction in variability does indeed seem to occur. However, as some of our graphs in Chapter 3 demonstrate, large differences in how long it takes to progress from task to task usually persist even in hierarchy-based instruction.

As research on the processes of learning and understanding mathematics proceeds, it seems reasonable to expect a better understanding of the individual differences that are so regularly observed in both the laboratory and the classroom. Cognitive task analyses, particularly if they focus directly on the processes of learning as well as on performance, should begin to reveal the kinds of knowledge structures that distinguish individuals who are successful in acquiring a new mathematical concept or skill from those who are less successful. They should also begin to characterize the learning processes of those who generally progress quickly through a mathematics curriculum as opposed to those who progress more slowly. As these processes are better understood, it should become possible to give a deeper psychological interpretation to the notion of mathematical aptitude.

Although batteries of aptitude tests usually include a "quantitative" or mathematical section, and although quantitative aptitude scores tend to predict rather well how students will do in learning mathematics and various sciences that demand mathematical competence, little is understood about why this relationship holds. The tests, it is true, depend heavily on acquired or learned knowledge of mathematics, but they also require certain reasoning and inference skills that are not directly taught in the mathematics curriculum. Detailed cognitive process analyses of performance required in taking aptitude tests, including the quantitative tests, are just now being undertaken by psychologists. (See Pellegrino & Glaser, 1980, for a review of this work.) These analyses, it may be hoped, will eventually allow us to describe test performance in terms of the same processes that the type of cognitive task analysis described in this book is uncovering for mathematics performance and learning. As we begin to understand better the kinds of processes involved in both learning the mathematics curriculum and taking quantitative tests, we should be in a position to discover what shared processes can account for the ability of the tests to predict school learning performance. At the same time, having a process-based definition of mathematical aptitude should make it possible to design instructional treatments that effectively adapt to individuals' characteristic modes of processing. In this way, the concept of aptitude-treatment interactions, now only a promise, may become a realistic way of making mathematics instruction effective for more people.

APPLYING PSYCHOLOGY TO INSTRUCTION

Our concern throughout this book has been to establish the framework for a dialogue between theory and practice. If the psychology of mathematics indeed follows the lines of questioning that we have outlined here, it seems likely that the issue of how to link psychological research with the practical concerns of instruction will largely melt away. As in Thorndike's psychology of arithmetic, applications seem to flow naturally from the kind of research we have envisaged for the future psychology of mathematics. Several aspects of our proposed agenda for the psychology of mathematics contribute to this prediction.

First, the research we envisage will focus squarely on the subject matter of mathematics. Those interested in applications to instruction will not have to search for how general principles of learning and thinking apply to a particular topic. Rather, it will be characteristic of the psychological research itself to have studied topics and tasks drawn directly from an existing or proposed mathematics curriculum. Curriculum designers and teachers will of course have to create practical materials and instructional patterns for the environment of the classroom. But they will not—as was the case, say, in applying behavioral theory or early structure-oriented theories to education—need to do most of the task analysis work on their own.

Second, by virtue of focusing their research efforts on tasks drawn from the curriculum, psychologists will largely avoid the problem of using laboratory tasks that are artifically simplified and disconnected from other knowledge. The tasks studied will be "messy" and poorly defined enough to match in these respects the tasks of the actual curriculum. For this reason, teachers and instructional designers will find less of a gap than in the past between the tasks studied by psychologists and the tasks they include in their curricula. They will thus be able to make more direct use of the psychologists' findings.

Third, psychological research will focus directly on the processes involved in learning and instruction, not only on describing successive levels of competence. Questions surrounding the organization of practice, the use of metaphorical representations, and the relationships between computation and understanding will be addressed directly. All this will mean, again, that those concerned with instruction will be able to draw quite directly on what psychologists are able to say about the nature of mathematics learning. Part of the task of translating from psychological findings to instructional programs will already have been accomplished.

Optimistically, perhaps, but not unrealistically, we look forward to a true psychology of instruction, a psychology that explicitly attempts to link processes of learning to the design and conduct of teaching. By exploring the relations between performance skill and conceptual understanding, the nature of cognitive representations and the role of instructional metaphors for abstract concepts, the role of practice in the growth of understanding, and the cognitive processes that underlie individual differences in learning, we think psychologists can begin to build a theory in which prescriptions for instruction are tied directly to knowledge about how learning proceeds in a complex intellectual domain. This new psychology of instruction will emerge as psychologists turn their attention to learning and performance in specific domains of knowledge. We expect the psychology of mathematics to be in the vanguard.

REFERENCES

Anderson, J. R. (Ed.). *Cognitive skills and their acquisition.* Hillsdale, N.J.: Lawrence Erlbaum Associates, in press.

Cronbach, L. J., & Snow, R. E. *Aptitudes and instructional methods: A handbook for research on interactions.* New York: Irvington, 1977.

Klahr, D., & Wallace, J. G. *Cognitive development: An information-processing view.* Hillsdale, N.J.: Lawrence Erlbaum Associates, 1976.

Neches, R., & Hayes, J. R. Progress toward a taxonomy of strategy transformations. In A. M. Lesgold, J. W. Pellegrino, D. Fokkema, & R. Glaser (Eds.), *Cognitive psychology and instruction.* New York: Plenum Press, 1978.

Pellegrino, J. W., & Glaser, R. Components of inductive reasoning. In R. E. Snow, P. A. Federico, & W. E. Montague (Eds.), *Aptitude, learning, and instruction: Cognitive process analyses* (Vol. 1). Hillsdale, N.J.: Lawrence Erlbaum Associates, 1980.

Thorndike, E. L. *The Thorndike arithmetics* (3 vols.). Chicago: Rand McNally, 1917-1924.

Author Index

Numbers in italics indicate the page on which the complete reference appears.

A

Almy, M., 187, *194*
Anderson, J. R., 198, *236*, 246, *253*
Anderson, M. C., 89, *95*
Anderson, R. C., 33, *37*, 89, *95*
Armstrong, R., 83, *96*
Austin, G. A., 111, *127*

B

Balfour, G., 49, *65*
Baylor, G. W., 182, *194*
Bearison, D. J., 180, 184, *194*
Beckwith, M., 60, *65*, 69, 70, 73, *95*
Behrens, M. S., 19, 26, *36*
Beilin, H., 187, *194*
Bell, E. T., 135, *153*
Blumberg, P., 225, *237*
Bobrow, D. G., 90, *95*
Boozer, R. F., 45, 62, *66*
Bower, G. H., 198, *236*
Brainerd, C. J., 176, *194*
Brian, D., 26, 28, *37*
Briggs, L. J., 41, 42, *65*
Brown, J. S., 87, 88, 93, *95*, 200, *236*
Brown, L. T., 146, *153*, 219, 220, *236*
Brownell, W. A., 17, 18, 19, 22, *36*
Bruner, J. S., 104, 111, 112, 113, 114, 115, 117, 119, *127*
Burton, R. R., 87, 88, 93, *95*, 200, *236*
Buswell, G. T., 24, 33, *36*

C

Carroll, J. B., 43, *65*
Caruso, J. L., 56, *65*
Case, R., 182, 183, 184, 188, *194*
Chase, W. G., 208, *236*
Chazal, C. B., 17, *36*
Chi, M. T. H., 73, *95*
Clapp, E. L., 20, 22, *36*
Collins, A. M., 198, 199, *236*
Craik, F. I. M., 31, *36*
Cronbach, L. J., 56, *65*, 145, *153*, 251, *253*

D

Dienes, Z. P., 55, *65*, 116, 117, 118, 119, 120, 121, 122, 123, *127*
Dilley, C. A., 107, *127*
Dolecki, P., 176, 180, 183, *195*
Donaldson, M., 49, *65*
Duckworth, E., 187, *194*
Duncker, K., 140, 141, *153*

E

Eggleston, V. H., 70, *96*
Elkind, D., 178, *194*
Ellis, H. C., 39, *65*

F

Flavell, J. H., 41, *65*, 144, *154*, 174, *194*

254

G

Gagné, R. M., 39, 40, 42, 47, 48, 49, 65, 146, *154*, 219, 220, *236*
Gallistel, C. R., 45, 59, 65, 69, 74, *95*
Garstens, H. L., 39, 40, 47, 48, 49, 65
Gascon, J., 182, *194*
Gattegno, C., 120, *126*
Geeslin, W. E., 208, *237*
Gelman, R., 45, 59, 65, 69, 74, *95*, 178, 179, 184, *194*
Ginsburg, H., 87, *96*
Glaser, R., 29, *36*, 50, 65, 225, 231, *237*, 252, *253*
Golding, E. W., 55, 65, 116, 118, 119, 120, 121, 122, *127*
Goodnow, J. J., 111, *127*
Greenfield, P. M., 111, *127*
Greeno, J. G., 124, *127*, 200, 201, 205, 206, 221, 224, *237*
Groen, G. J., 74, 75, 78, 79, 81, 83, *96*, 186, *194*
Guttman, I., 44, 65

H

Hartley, J. R., 33, *37*
Hayes, J. R., 247, *253*
Hightower, R., 106, *127*
Hogaboam, T., 33, *37*
Holt, M., 121, *127*
Huey, E. B., 30, *36*
Hunt, J. M., 189, 192, *194*
Hydle, L. L., 22, *36*

I

Inhelder, B., 157, 159, 161, 168, 170, 171, 172, 173, 188, *194*, *195*
Isen, A. M., 176, 180, 183, *195*

J

Jackson, A. E., 107, *127*
Jacobson, E., 29, 33, *36*
Jerman, M., 23, 26, 28, *36*, *37*
Judd, C. H., 39, *65*
Judd, W. A., 29, *36*

K

Kagan, J., 187, *194*
Kaplan, J., 46, 50, 51, 52, 53, 58, 59, 61, 63, 65

Katona, G., 142, 143, 144, *154*, 219, *237*
Keislar, E. R., 145, *154*
Klahr, D., 70, 73, *95*, *96*, 182, *195*, 247, *253*
Knight, F. B., 19, 20, 26, *36*, *37*
Koffka, K., 103, *127*
Köhler, W., 103, *127*, 131, 147, *154*
Kramer, G. A., 22, *36*
Kresh, E., 56, *65*
Kreutzer, M. A., 144, *154*

L

LaBerge, D., 33, *36*
Landauer, T. K., 76, 83, *96*
Lankford, F. G., 84, 85, 88, *96*
Larkin, J. H., 207, 224, *237*
Laudato, N. C., 91, 94, *96*
Leonard, Sister C., 144, *154*
Lesgold, A. M., 33, *37*
Liebert, D. E., 188, *195*
Liebert, R. M., 108, *195*
Lockhart, R. S., 31, *36*
Loftus, E. F., 22, *36*
Lord, F., 44, *65*
Lovell, K., 190, 192, *195*
Luchins, A. S., 129, *154*
Luchins, E. H., 129, *154*

M

Magone, M. E., 219, 224, *237*
Maier, N. R. F., 218, *237*
Mayor, J. R., 39, 40, 47, 48, 49, *65*
McConnell, T. M., 18, *36*
McLanahan, A. G., 176, 180, 183, *195*
Miller, G. A., 31, *37*
Montessori, M., 108, *127*
Morningstar, M., 26, 29, *37*, 86, *96*
Morris, L. L., 225, 230, *237*
Moyer, R. A., 76, 83, *96*
Mulholland, T., 225, *237*
Murray, F. B., 192, *195*

N

Neches, R., 247, *253*
Newell, A., 220, 230, *237*
Norem, G. M., 20, *37*
Norman, D. A., 198, *237*
Novick, M., 44, *65*

O

Olver, R. R., 111, *127*

P

Paige, J. M., 90, 92, 93, 94, *96*
Palermo, D. S., 49, *65*
Paradise, N. E., 39, 40, 47, 48, 49, *65*
Parkman, J. M., 74, 75, 78, 83, *96*
Pascual-Leone, J., 183, 184, *195*
Payne, J. N., 106, *127*
Pellegrino, J. W., 225, *237*, 252, *254*
Perfetti, C. A., 33, *37*
Piaget, J., 103, *127*, 157, 159, 161, 165, 166, 168, 170, 171, 172, 173, 178, 188, 190, *194, 195*
Poll, M., 186, *194*
Polya, G., 149, 150, 152, *154*, 234, *237*
Prince, J. R., 187, *195*

Q

Quillian, M. R., 198, 199, *236*

R

Rabinowitz, R., 187, *194*
Rasmussen, L., 106, *127*
Rasmussen, P., 106, *127*
Rathmell, E. C., 106, *127*
Rees, R., 23, *36*
Reitman, J. S., 208, *237*
Repp, A. C., 24, 25, 26, *37*
Resnick, L. B., 44, 45, 46, 47, 50, 51, 52, 53, 56, 58, 59, 61, 62, 63, *65, 66*, 79, 81, 83, *96*, 225, 231, *237*
Restle, F., 60, *65,* 69, 70, 73, 83, *95, 96*
Riley, C. A., 176, 180, 183, *195*
Roman, R. A., 91, 94, *96*
Rothkopf, E. A., 209, *237*
Rubin, E., 83, *96*
Rucker, W. E., 107, *127*
Rumelhart, D. E., 198, *237*

S

Samuels, S. J., 33, *36*
Schadler, M., 225, *237*
Schaeffer, B., 70, *96*

Scott, J. L., 70, *96*
Sekuler, R., 83, *96*
Shavelson, R. J., 208, 209, *237*
Shulman, L. S., 145, *154*
Siegel, A. W., 56, *65*
Siegler, R. S., 188, *195*
Simon, H. A., 90, 92, 93, 94, *96*, 208, 220, *236, 237*
Sinclair, H., 178, *195*
Snow, R. E., 56, *65*, 251, *253*
Spiro, R. J., 89, *95*
Stanton, G. C., 208, *237*
Stretch, L. B., 22, *36*
Suppes, P., 22, 26, 28, 29, *36, 37,* 74, 83, 86, *96*
Swenson, E. J., 18, *37*
Szeminska, A., 157, 159, 161, *195*

T

Tait, K., 33, *37*
Thorndike, E. L., 12, 13, 14, 15, 16, *37*, 39, 65, *148, 251*
Thio, M. P., 209, *237*
Thurner, R. D., 209, *237*
Trabasso, T., 176, 180, 183, *195*
Trafton, P., 102, *127*
Tucker, T., 176, 180, 183, *195*

U

Underwood, B. J., 62, *65*
Uprichard, A. E., 49, *65*

W

Wallace, J. G., 70, 73, *96*, 182, *195*, 247, *253*
Wang, M. C., 45, 46, 50, 51, 52, 53, 58, 59, 61, 62, 63, *66, 67*
Wertheimer, M., 130, 132, 133, 135, 136, 139, *154*, 218, *237*
Wheeler, L. R., 17, 20, 22, 26, *37*
White, R. W., 44, *66*
Wickelgren, W. A., 234, 235, *237*
Woods, S. S., 79, 81, *96*
Woodworth, R. S., 39, *65*

Y

Young, R. M., 182, *195*

Subject Index

A

Ability, *see also* Individual differences; Thinking, quantitative, 234
Abstraction, *see* Generalization; Symbolic representation
Accuracy, 24
  and automaticity, 33
  and CAI, 26
  as a criterion of proficiency, 11–12, 17, 19, 35, 97
  and individual differences, 29
  as a measure of difficulty, 38
Achievement tests, *see also* Evaluation, of student; Stanford Achievement Test
Active processing, *see also* Information processing psychology, 198
Active thinking, 116, 129, 164–165, 191, 193, 200, 231, 249–250
Adaptation, teacher and child, *see also* Development, and optimizing instruction, 55
Addition
  and AI, 92
  and CAI, 26
  in columns, 14
  common addend problems, 20
  as computation, 10
  and drill, 11
  errors, 84–87
  and learning hierarchies, 39–41, 47–48
  missing addend problem, 185
  and problem difficulty, 19–22
  and problem solving, 215
  process models, 68, 74–82
  and physical materials, 249
  and representation, 203–205
  reversal of pairs problem, 20
  and underlying logic, 17–19, 196, 247
Algebra, 105, 122, 164, 221
  as a complex subject, 7, 38
  and word problems, 68, 90–94, 95, 106, 196, 214–217
Algorithms, 10, 31–32
  and automaticity, 31–33, 35
  blind, 133–135
  errors in, 83–88, 95, 210–213
  and problem solving, 68, 214, 220
  and underlying structure, 106, 125, 196, 210–213, 249
  "all or none" concept, 174
  "all other things being equal", 171
American question, 178
Analysis, *see also* Goal analysis, 141–142
  and problem solving, 116
  of task, *see* Empirical task analysis; Rational task analysis; Task analysis
Angles of a triangle problem, 157–161
Anti-Piagetian theory, 155, 174–182
Apparent motion, 130
Applied mathematics, *see also* Contextual effects; Meaningfulness, 89

257

258 Subject Index

Aptitude, *see also* Individual differences; Thinking, quantitative, 252
Aptitude treatment interactions, 27, 56, 145, 251-252
Area, *see also* Carpenter's apprentice problem; Parallelogram problem; Spatial representations; Tightrope problem, 132-136, 149-151, 181, 225-234, 236
Arithmetic, 10, 74-89
  and associated strategies, 11, 95, 247
  and conceptual understanding, *see also* Understanding, 196
  and development, *see also* Development, 168, 189
  as focus of mathematical curriculum, 7, 11, 35
  and materials, *see also* Materials, 106-110
  as method of expression, 122
  as part of mathematics, 105
Artificial intelligence, *see also* Computer simulation, 90
  and human problem solving, 92
Association method, *see also* Correspondence, 208-209
Associationism, 5, 7, 111, 152, 177, 243, 245, 253
  and instruction, 12-17, 24-26, 35, 38-39, 147
  and knowledge, 198, 202-203, 250
  problem difficulty, 19-20, 38
  and transfer, 38
  and understanding, 128
Associations, *see* Bonds
Associative law, 117, 189
Attention
  directed, 179-180, 184, 233-234
  and environment, 175
  and information, 31
Attribute blocks, 119-120
Automaticity, *see also* Algorithms; Factual knowledge, 30-35
  and drill and practice 17-18
  and working memory, 62, 247
Axioms, *see also* Mathematical principles, 105

B

Balance beam, 115, 119
Banana problem, 132, 152
Base representation, 106-110, 115-119, 122, 210-213, 248

Behavioral analysis *see* Associationism
Behavioral data, *see* Performance data
Behavioral objectives *see* Instruction, goals and contents of
Bending rods problem, 168-173
Blocks; *see* Dienes blocks, Materials
Board problem, 93, 151
Bonds, 12
  as content of learning, 17-19, 34, 35, 124, 128, 242
  formation of, 13-15
  and learning hierarchies, 39, 41, 64
  organization of, *see also* As semantic links, 15, 128, 250
  propadeutic, 14
  selection of, 13-15
  as semantic links, 199-205, 214, 230
Borrowing, 34, 41, 210-213, 249
Bottom up processing, *see* Top down versus bottom up processing
Buggy algorithm, 210
Bugs, *see also* Errors, 87-88
Burst, 207

C

CAI, *see* Computer assisted instruction
Calculating machine, 249
Calculations, *see* Computation
Calculus, 105, 164
Cambridge conference, 103, 128, 147
Card sorting tasks, 208
Cardinal quantification, 45, 74
Carpenter's apprentice problem, 135-138, 147, 151
Carrying, 14, 106, 110, 115, 117, 210, 213, 249
Checking, 86
Chess, 208
Chimpanzee problem, 131
Choice model, *see* Subtraction, choice model
Chunks, 31, 70, 73, 95, 183, 208
Classification tasks, 111, 119-120, 166-168, 172, 175, 176-177, 180, 182, 221-225
Clinical interviewing, 156, 157, 206, 210
Closure, *see also* Gestalt psychology, 130-131, 152
Clustering, 209
Cognitive abilities, *see* Individual differences
Cognitive demand, *see* Memory load; Working memory
Cognitive objectives, *see* Instruction, goals and contents of

Cognitive processes, *see also* Cognitive psychology, 39, 103, 111, 253
Cognitive psychology, *see also* Cognitive processes; Information processing psychology, 6, 7, 99, 124, 243-244, 248
Cognitive task analysis, 251
College, 103
Columns, *see* Formatting conventions
Commutativity, 9, 20, 76, 117, 189, 203, 206
Comparison, *see also* Sets, 60-61, 166, 181
Competence, *see* Proficiency
Competence versus performance, 175-176
Complex intellectual processes
  as higher order problem solving skills, 38, 69, 84, 89, 97, 102, 148
  as a sequence of operations, 10, 41, 55-56, 91, 205
  versus simple laboratory tasks, 3, 5, 241
Complex-first learning, 56
Component skills, *see* Learning hierarchies; Skills
Computation
  and arithmetic, 7, 9-10, 11, 39, 97-98, 102, 106, 124-125
  as final step, 151
  and problem difficulty, 38
  procedures of, 17, 56, 63, 67-68, 88-89
  processing demands, 12, 32
  and understanding, 83, 101, 196-197, 200, 246, 249, 250, 253
Computer assisted instruction, 12, 26-30, 35, 86-89, 94
  and problem difficulty, 22, 26-30, 35
Computer simulation, *see also* STUDENT; PERDIX, 68, 90, 94-95, 182, 197, 230-231
Concept blocks, 27
Concept oriented materials, *see also* Materials, 116-120
Concepts, *see also* Mathematical principles
  and general problem solving strategies, 221, 225
  versus computation, 7, 111, 125, 126, 197
  versus procedures, 210, 250, 253
Conceptual approach, *see also* Mathematical principles; Structure oriented teaching, 101-102
Concrete experiences, *see* Materials
Concrete materials, *see* Materials
Concrete operations, 113, 165, 167-168, 174, 187, 191, 193-194
Condition-action statements, 230
Conflict, 132, 148, 183-184, 189, 192

Conjunction, 120
Connectedness, 206, 210-213, 235
Connectionism, *see* Associationism
Connections, *see* Bonds
Conservation, 73-74, 166-168, 175, 178-182, 183-184, 189, 192
Construction, *see also* Active thinking; Structure, 164-165, 190-191, 194, 200
Constructivism, *see* Active thinking
Contextual effects
  environmental, 129-131, 191, 234
  and generalization, 122, 124
  mathematical, 55-56, 103-105, 106, 138
  real life, 5, 89, 94, 95, 103, 253
Contiguity, 199-200
Continuous quantity, 181
Coordinate systems, *see* Graphs
Coordination of multiple dimensions, 162, 168
Correctness, 206-207
Correspondence, 208-209, 217, 225, 235
Count fixed unordered arrays, 59-60, 62
Count movable objects task, 58-59, 61-62
Count subset task, 60-61
"Counter", 75
Counting, 58-62, 68, 69-74, 95, 175, 215
  and addition, 22, 74-82, 85, 87, 112, 186
  fingers, 112, 117
  independence from one to one correspondence, 45-46
Cramming, 208
Creativity, *see also* Invention, 19
Criterion referenced tests, 50
Cuisinaire rods, 120, 121
Cumulative learning theory, 39, 43, 64, 205, 207, 243
  instructional implications, 50, 55
Curricula
  development, 23, 38-39, 47, 50, 54, 57, 58, 62-64, 187, 243-244, 245
  goals of, 11, 13, 103-104, 123, 251-252
  reform, 7, 9, 33, 101-105, 111, 123, 124, 248

D

Decimal system, *see* Base representation
Decimals, 14, 101
Decrementing model, *see* Subtraction, decrementing model
Deep end approach, 55-56
Definitions, formal, *see also* Mathematical principles, 209, 225

260 Subject Index

Development
  and optimizing instruction, 4, 7, 20, 23, 101–102, 104, 111, 113, 123, 126, 177, 182–186, 186–193
  and performance differences, 70, 73–74, 79, 217
  and the environment, 99
  as a general concern of psychology, 3, 244
  fixed order, 45
  sequence of, 41, 45, 69, 112–114, 126, 155, 165–174, 187, 190, 193
Diagnosis, 63–64, 84–89
Diagrams, see Spatial representations
Dienes blocks, 117–119, 122, 210–212, 248–249
Difficulty, see Problems, difficulty of; Skills, difficulty of
Discontinuous quantity, 181
Discovery
  enhancing, 234
  examples, 141
  guided, 145–146, 219–220
  imitations of, 183
  learning, 144–146, 148, 236
  motivation for, 7, 85, 105, 130, 152, 260
Disjunction, 120
Distractors, 116
Distributive law, 117
Division, 10, 11, 84, 85, 92, 201–203, 219
  long, 10, 15
Domain specificity, 175
Door problem, 140
Dot problem
  Gestalt, 129–131
  Piagetian, 162–164
Drill and practice
  and automaticity, 18
  and CAI, see computer assisted instruction
  description of, 11
  and learning, 247–248
  limitations of, 17–18, 34, 102, 105, 250
  mechanism, 13
  optimizing, 4, 15, 18–19, 20, 24–30, 33, 35, 38, 55, 124–125, 126, 246–247, 253
  spaced versus mass, 24, 35
Dynamism, see Active thinking

E

Elegance, 135, 138, 147–148
Empirical task analysis, 58, 63, 64, 67–83, 91, 98

Enactive representation, 112–113, 115, 122, 126, 147, 214
Encoding, 175–180, 214–217
Environment, see also Functional fixedness
  as a source of information, 30
  as determinant of learning, 156, 177–178, 187
  social, 156, 174, 191–192, 194
  task, see Task environment
Equations, 91, 93, 95, 113, 186, 190, 209, 216, 221
Equilibrium, 131
Equivalence, 202
Errors
  avoidance of, 26–27, 138
  and automaticity, 32
  correction, 25, 26–27, 121, 210–213
  and frustration, 95
  and problem difficulty, 20
  as a source of information, 4, 68, 83–89, 182, 192, 196
Estimation, 68, 162, 164
Evaluation
  of student, 36, 61, 68, 80, 115, 124–125, 126, 175–177, 182, 197, 205–213
  of theory, see also Methodology, 43–57, 63, 64, 123–125, 135, 148, 153, 180, 192, 207
  self, 235
Evolution, 156
Excitement, see also Motivation, 104
Expanded notation, 212
Expert-novice comparison, 82, 207–209, 224
Experts, 78

F

Factoring principle, 117, 119
Factual knowledge, 10, 17, 31, 41, 68, 74, 83–84, 88, 92, 217, 225
Fading out, 212
Feedback, 12–15, 119, 123, 180, 191, 194, 243
  in CAI, 26, 27
Fingers, counting on, 17, 19, 85, 112
Flash cards, 11
Flexibility, 19, 235
Flow charts, 58–59
Forgetting, see also Recall, 61–62, 90, 198
Formal operations, 113, 168–174, 188, 193, 194
Formatting conventions, see also Base representation, 68, 113, 186, 210
Formulae, 146
Fractions, 10, 14, 32, 84, 85, 120, 124

Free play, 120, 234
Free recall, 207
Function theory, 7, 9, 105
Functional fixedness, 217-219, 236
Functional theory, 182

G

Games, 121
Gap-extra relationship, *see also* Parallelogram problem, 134-135, 232-234, 236
General problem solving strategies, 234-235
Generalization, 121-122, 158, 161, 182, 198-201, 249
Generate and test strategy, 223, 236
Geography, 43, 115
Geometry, *see also* Parallelogram problem; Parallel line problem, 3, 7, 9, 94, 104, 105-106, 122, 134-135, 149-151, 156, 168, 190, 221-224, 232-234, 236, 250
Gestalt psychology, 7, 99, 115, 155, 200, 214, 218, 244, 248, 250
 and instruction, 139, 145, 146-152
 principles of, 129-137
 productive thinking, 132, 138-146, 147-148
Go, 208
Goal analysis, 67, 148, 149, 151, 217-218, 231-234, 236
Goal oriented trials *see* Problem solving, goal oriented trials
Goals, *see also* Goal analysis, 223-224, 232
Good inner relatedness, 148
Grammar, *see also* Problems, difficulty of; Structural variables, 23, 89
Graphs, 43, 122, 164
Grouping, 60, 70, 108, 110, 129-131
Guessing, *see also* Estimation, 17, 233
Guttman scale, 44, 64

H

Heuristics, 68, 149, 223, 235, 236
Hierarchical properties, 199-201
Hierarchical task analysis, *see* Learning hierarchies
Higher level tasks, *see also* Learning hierarchies; Skills, and subskills, 41-43
Hints
 as environmental cues, 217, 218-219, 236
 as experimental procedure, 141-146, 218, 228-231
 and instruction, 94, 148-149, 153, 174, 225

Horizontal decalage, 174-175
Human information processing, *see* Information processing psychology
Hypothesis information, 161

I

Iconic memory, *see* Sensory intake register
Iconic representation, 112, 115, 122, 126, 147, 214
Identical elements theory, *see also* Transfer; Cumulative learning theory, 38-39
Imagery, *see also* spatial representations, 112, 115
Impossible problem, *see* Board problem
Incentive, *see* Motivation
Incidental learning, 55-56
Incrementing model, *see* Subtraction, incrementing model
Indirect problem solving, *see* Problem solving, detours in
Individual differences, 250-252
 and CAI, 26-30
 and instruction, *see also* Matching, 39, 50-57, 62-64, 98, 115, 121, 127, 145, 149, 187-189, 253
 specification of, 86, 88, 217, 236
Individual performance, 68, 88
Inequalities, 9, 83, 101
Inferences, 198, 200, 223, 236, 250, 252
Information processing psychology, 7, 30, 63, 68, 99, 115-116, 174, 196-236, 243
Inherency, 156, 174
Insight, 131-132, 135, 139, 146-152, 153, 155, 205, 218, 250
Institute for Mathematical Studies in the Social Sciences, 26
Instruction
 and associationism, 12-17, 38-39
 and Gestalt psychology, 139, 145, 146-152
 goals and contents of, 7, 9, 18, 39, 68, 74, 82, 95, 101-108, 113-115, 123-125, 126-127, 187-188, 189, 192, 193, 194, 205-207, 208, 234-235, 242-244, 245, 248, 253
 and learning hierarchies, 49-57, 62, 97, 251
 methods, 18, 105
 and Piagetian theory, 174, 182, 186-193
 sequence of, *see* Learning hierarchies
 structure oriented approach, *see* Structure oriented teaching
Integers, *see also* Numeral, 39, 41, 47

Integration, 206-208, 225, 235, 250
Intellectual activity, see Active thinking
Intellectual challenge, see Motivation
Intellectual functioning theory, 111
Intellectual recapitulation, 156
Intellectual skills, see Skills
Intelligence, see Individual differences; Thinking, quantitative
Intermediate term memory, see Working memory
Internal counting, 69, 73
Interpretation, see Active thinking
Intuition, 83, 102, 104, 120, 124
Invention, 63, 81, 82-83, 88, 95, 190, 228, 231-234, 247, 250
Inverse relationships, see also Reversibility, 201-205, 206

K

Knowledge, 31, 124
Knowledge structures, 99, 102, 197, 198, 201-205, 225-230, 235, 241, 250-257
  assessment of, 205, 213, 247
  cumulation of, 57, 230, 234
  definitional, 91, 221, 225
  organization of, 18, 128, 198, 205, 213
  problem solving, 4, 97, 148, 183
  procedural, 225

L

Labelled links, 199, 201-202
Language understanding, see Linguistic skills
Law of effect, see Principles of reinforcement
Law, political, 115
Learning, see also Discovery; Incidental learning; Invention; Learning hierarchies, 3, 5, 38, 42, 50, 68, 88-89, 103, 111, 201, 225-235, 236, 242-244, 246-250, 253
  cycle of, 120-123, 126, 219-220, 246-247
Learning hierarchies, 10, 39-49, 205, 243
  and instruction, 42, 49-57, 58, 62, 97, 251
  operation of, 39
Linguistic skills, 89-91, 124, 171, 176-177, 197, 198-201, 214-217, 241
Logic, 9, 99, 117, 119, 155, 172-177, 201
Long term memory, 5, 93, 198-200, 220
Longitudinal studies, 248
Lower level tasks, see Skills; Learning hierarchies

M

Manipulatives, see Materials
Mapping, 190
Mastery, see Proficiency
Matching, 189-193, 194, 251
Materials, 17, 47, 92, 101-102, 126, 145, 151-152, 175, 191, 207, 218, 248-249, 253
  blocks, 81, 116-122, 210-212, 217, 248
  Montessori, 108-110, 248
  sticks, 106-108
Mathematical ability, see Individual differences; Thinking, quantitative
Mathematical honesty, 234
Mathematical principles, 7, 9, 11, 17-19, 35, 39, 76, 95, 97-98, 101-105, 116-117, 122-123, 135, 137-138, 141-148, 151, 153, 180, 196, 200, 207, 208-209
Mathematical thinking, see Thinking, quantitative
Mathematical variability, 122, 126
Mathematics, 6, 7, 9, 36, 201, 241, 252
Meaningful learning, see Learning
Meaningfulness, see also Context; Learning; Mathematical principles; Structure oriented teaching
  of instruction, 18, 19, 34, 35, 102, 128
  of problems, 5, 15, 18, 94, 146
  versus unfamiliarity, 22, 101-102, 124
Measurement, 162-164, 189, 228-231
Mechanical solutions, 142
Memorization, see also Learning, rote versus meaningful, 30, 45, 122, 143, 144
Memory, see also Information processing psychology; Long term memory; Semantic memory; Sensory intake register; Working memory, 68, 112, 197, 227
Memory load, see also M-space; Working memory, 30-32, 34, 35, 70, 215-216
Mental actions, see Active thinking
Methodology, see also Evaluation 4-5, 7, 44, 47-49, 64, 67-68, 69, 76-78, 84, 90, 92, 94-95, 123, 150, 157, 182, 197, 206, 207-213, 228-230, 248
Missing addend problem, 185-186
Mixed numbers, see also Fractions, 14
Mnemonics, 208
Money, 14, 17, 93
Monkey problems, see Banana problem; Chimpanzee problem
Montessori, 108-109, 248

Subject Index   263

Motivation, 35, 105, 126, 183–184
Movement, 156
Movie analogy, 245
M-space, 183–184
Multibase arithmetic blocks, *see* Dienes blocks; Materials
Multiple connectedness, 203
Multiple embodiments, 121
Multiples of 13 problem, 141–142
Multiplication, 9, 10, 11, 14, 20, 70, 84, 85, 86, 87, 92, 201–205, 219
Multiplicative classification, 172
Music, 130

N

Natural language understanding, *see* Linguistic skills
Negation, 120
Neo-associationism, 200
Neo-Piagetian theory, 155, 182–186
Network theory, 197, 201, 205, 207, 213, 235
Neurological patterns, 147
Newtonian mechanics, 209
Newton's law of moments, 115
Nodes, 200–201
Nonsense syllables, 5
Notation, standard, 108–110, 113, 115, 119
Novice-expert comparison, *see* Expert-novice comparisons
Number facts, *see* Factual knowledge
Number lines, 94, 203
Number series problem, 143, 219–220
Numerals, *see also* Integers, 46, 58, 113, 120

O

Observation, *see* Performance data
Once and only once criterion, 59–60
One-to-one correspondence, 45–47
Operational thinking, *see* Concrete operations
Operations, 7, 23, 89, 105, 113, 186, 202–203, 206–207, 247, 248
  reversibility of, 83, 165, 203
Ordinal quantification, 45
Organic solutions, 142
Organization
  and learning, 143–144, 147
  and problem solving, 218
  of a spatial display, 129–131, 134
  of knowledge, 99, 205–207, 214

P

Parallel line problem, 221–224
Parallelogram problem, 132–134, 225–234
Partitioning, 91
Part-whole relationships, 99, 152–153, 155, 160–162, 177, 196, 218
  function of parts, 132, 134, 138, 149
  summation of parts, 129, 132
Pendulum problem, 188
Percents, 10
Perception, *see also* Part-whole relationships, 68, 70, 95, 99, 126, 129, 134–135, 147, 160–166, 180, 183, 236
Perceptual variability, 121–122
PERDIX, 221–224, 250
Performance capability, 41
Performance data, *see also* Methodology, 3, 58, 68, 83, 84, 95, 97, 176, 192, 196–197, 205, 207, 210, 243, 244–245, 250–251
Performance models, 68
Phenomenological reports, 129
Physics, 115, 209, 224, 241
Piagetian theory, 7, 111–112, 115, 155, 200, 244
  critiques of 174–186
  and development, 165–174, 193
  and instruction, 177–186, 186–193, 193–194
  thinking as structuring, 156–165
Place value concept, *see also* Base representation, 107, 120, 213
Pointer, 199
Pouring, 175
Practical exercises, *see* Context effects; Meaningfulness
Practice, *see* Drill and practice
Preoperational thinking, 167, 174, 178, 193
Prerequisite skills, *see* Learning hierarchies; Skills
Prime number problem, 223
Prime numbers, 140
Principles of reinforcement, 12–13
Problem solving, 10, 68, 97, 125, 205, 213–217, 236, 248
  and artificial intelligence, 92, 94
  as a concern of psychology, 6
  as a definition of mathematics, 7
  detours in, 131, 152
  and dynamic mental forces, 132
  errors in, *see* Errors
  from above, below, *see* Top down versus bottom up processing

goal oriented trials, 131
rule bound, 134-135
and similar problems, 149, 153
stages of, 149-150, 234-235
Problem space, 220
Problems, *see also* Word problems, 11
analysis of, 149, 214
definition of, 148, 220-221
difficulty of, *see also* Structural variables, 12, 15, 19-24, 35, 38, 87, 89, 210, 215-216, 219, 242, 248
and drill, *see* Drill and practice
knowns and unknowns, 148, 153, 215, 217, 234
Procedural knowledge, 87, 97, 147, 205, 210, 225, 249-250
Process analysis, *see also* Empirical task analysis; Rationale task analysis, 39, 41, 58, 182, 205
Procedural networks, 88, 200
Procedures, *see also* Operations; Procedural knowledge, 23-24, 84, 205, 217, 220, 246-247, 249
selection of, 139
Processing, *see* Information processing
Production systems, 230
Productive thinking, *see* Gestalt psychology, productive thinking
Programmed instruction movement, 243
Prompts, *see* Hints
Proofs, 102, 105, 171, 221-225
Properties, 199
Propositions, 124
Protocols, *see* Clinical interviews; Phenomenological reports; Thinking aloud protocols
Psychological equilibrium, *see* Closure; Gestalt psychology
Psychological field, 99, 130, 132
Psychological universals, 4
Psychology
of arithmetic, 13, 16
of mathematics instruction, 3, 4, 6-7, 11, 90
Psychology of subject matter, 4-6
Puzzle box experiment, 12

Q

Quadratic equations, 113, 117
Qualitative analysis, 224
Quantification, *see also* Counting; Subitizing, 165-168, 180, 248

Quantitative thinking, *see* Thinking, quantitative

R

Ratio, 120
Rational task analysis, 39, 58, 62-64, 67, 90, 182, 205, 206, 241
Reaction time, 69, 71-73, 78-79, 83-84, 94, 197, 200, 203, 207-208, 248
Readiness, *see* Development, and optimizing instruction; Matching
Reading, 5, 33, 89
Real life tasks, *see* Contextual effects
Reasoning, *see also* Thinking, 4, 140
Recall, 198, 207, 213
substitute procedures, 22, 30, 74, 198
Recognition, *see also* Insight, 40, 67, 92, 132, 140, 155
Reductionism, *see also* Associationism, 129
Reformulation, *see* Reorganization
Rehearsal, 31
Reinforcement theory, *see* Associationism
Relatedness coefficient, 208-209
Relations, *see* Labelled links
Remembering, *see also* Recall, 5, 17, 126, 198, 207, 213
Reorganization, 105, 111, 132, 140, 143-144, 149, 151, 191, 247
Representation, 102, 110-116, 147, 191, 196, 210, 248-249
and problem solving, 94, 95, 153, 197, 214-220, 235-236
Research questions, 4-6, 34, 64, 88-89, 94, 95, 98, 110, 115, 125, 126, 148, 153, 194, 207, 235, 247
Response flexibility, 19, 235
Restructuring, *see* Reorganization
Retention, 142-143, 146, 180-182, 187, 250
Reversibility, 165-167, 174-175, 193
Reward, *see* Feedback
Rote memory, *see* Memorization
Rounding procedures, 68
Rules, *see* Condition action statements

S

Scaling studies, 43-47, 63, 64
Schemes, 182-186
School, 3, 6, 9, 11, 35, 38-39, 50, 54, 97, 111, 125, 175, 192, 194, 241

Search, of problem solving space, 140, 213, 223, 235-236
Semantic analysis, 197, 201-205
Semantic memory, 197, 198-201, 235
Semantic networks, *see* Network theory
Senseless learning, *see* Learning, rote versus meaningful
Sensorimotor stage, 112, 174
Sensory buffer, *see* Sensory intake register
Sensory intake register, 30
Sequencing
  of operators within a problem, 23
  of problem solving events, 140
  of skill acquisition, *see also* Development, and optimizing instruction; Matching, 47-49, 50, 55-57, 62, 97
Seriation tasks, 112, 182
Sets, 7, 9, 18, 45-46, 49, 58-59, 60-62, 70, 74, 82, 101, 105, 117-120, 168, 176-177
Short term memory, *see* Working memory
Simplification, 138, 183-205
Simultaneous equations, 106
Skills, 3, 6, 54, 68, 89, 102, 138, 246-247, 248, 252, 253
  difficulty of, *see also* Errors, 62-63
  and computation, 97, 124-125
  and practice, 24
  and subskills, *see also* Learning hierarchies, 39, 41, 57, 64, 175, 205
Sorting tasks, 111
Space age, 102
Space race, 102
Spatial ability, 58
Spatial representations, *see also* Carpenter's apprentice problem; Materials parallelogram problem; Tightrope problem, 60, 63, 70-71, 93-94, 107, 112-115, 122, 136, 138, 151, 156-165, 166-168, 190, 216-217, 220
Speed, *see also* Automaticity; Reaction time, 11, 12, 19, 24, 26, 30, 33, 35, 97, 200
Spiral curriculum, 34, 104, 120-121, 125, 147
Sputnik, 102
Square roots, 34, 67-68, 217
Squaring, 92
Stages, *see* Piagetian theory, and development
Staircase problem, *see* Carpenter's apprentice problem
Stanford achievement test, 29
Statistical procedures, 57, 64, 79
Sticks, *see* Materials
Stimulus response learning, *see* Associationism

Story comprehension, 89
Story problems, *see* Word problems
Strategy, *see also* Subgoals, 68, 85, 89, 94, 95, 139-140, 148, 153, 183-184, 186, 197, 213-214, 216-217, 221-225, 230, 234-235, 247
Structural variables, *see also* Problem difficulty, 22, 35
Structure
  of general problems, 131-132, 144-146, 152
  and knowledge, 124, 197, 198-200, 206, 233-234, 235
  mathematical, 82-83, 88, 98, 102, 103-105, 128, 135, 137-138, 143, 147, 149, 151
  mathematical versus spatial, 138
  of subject matter, 4, 82, 147-148, 165, 196, 206, 234
  interference with, 145, 146
  Piagetian, 156-157, 164-165, 174-177, 182
Structure oriented teaching, 102, 105-110, 116, 123-125, 126-127, 128, 147, 190, 249
Subgoals, 223, 225, 230, 232, 235, 236
Subitizing, 70-72, 74, 95, 184
Subskills, *see* Skills
Subtraction, 41-42, 68, 74, 79-81, 83-85, 92, 95, 186, 196, 203, 210-213, 219
  choice model, 79, 83
  decrementing model, 79
  difficulty, 20
  incrementing model, 70, 203
Successive approximation, 243
Symbolic representation, 113, 117, 120, 193, 214, 218, 249
  and instruction, 105-110, 113-115, 122-123, 126
  translation process, 91-94, 95, 216-217
Symmetry, 117
Syntax, *see* Formatting conventions

T

Task analysis, *see also* Cognitive task analysis; Empirical task analysis; Learning hierarchies; Process analysis; Rational task analysis, 10, 205, 230, 242, 245
Task environment, 197, 214, 217-220, 230, 234, 235-236
Temporal patterns, *see* Reaction time
Tests, *see* Evaluation
Textbooks, 20, 47, 85, 107, 117, 124, 209, 242
Thinking, 3, 4, 31, 67-68, 102, 111, 130, 156-157

quantitative, 3, 7, 17–18, 34, 35, 99, 125, 188, 252
Thinking aloud protocols, 84, 90, 92, 94–95, 123, 156, 197, 221
Tightrope problem, 149–151
Time, 156
Top down versus bottom up processing, 139–140, 153
Training studies, 47–49, 63–64, 183
Transfer, see Cumulative learning theory; Generalization
Transformations, mental, see also Reorganization, 166, 175, 225, 228
Transitions in competence, 41–42, 82, 182, 244
Transitivity, 93
Trapezium, 160
Triangle problem, 157–162, 168
Trick questions, 218–219
Tumor problem, 140
Two string problem, 218–219, 224

U

Understanding, 18, 30, 56, 76, 94, 99, 101, 110, 126, 190–191, 196–197, 205, 209, 213, 235, 244–248, 250, 253
Unifix cubes, 121
Units, conversion of, 23
Universals, see Psychological universals

Unknowns, see Goal analysis; Problems, knowns and unknowns

V

Variability, see Mathematical variability
Variables, 169–173
Venn diagrams, 119–120
Verbal instruction, 63, 217, 219–220
Verbalization, see also Thinking aloud protocols, 123, 191
Visual representation, see Spatial representation

W

Well structured knowledge, 205–213, 225–230, 235–236
Wholistic perception, see Part-whole relationships
Woods Hole Conference, 103–104, 128, 147
Word associations, 206, 208
Word problems, 10, 89–94, 95, 198, 210, 215–216
Word recognition, see Reading; Linguistic skills
Working backwards, 235
Working memory, 31–32, 58, 60, 62, 95, 198, 247

X,Y,Z

Zero, 14, 101, 212